Electronic Message Transfer and Its Implications

Electronic Message Transfer and Its Implications

Alfred M. Lee
Cornell University

LexingtonBooks
D.C. Heath and Company
Lexington, Massachusetts
Toronto

Library of Congress Cataloging in Publication Data
Lee, Alfred M.
Electronic message transfer and its implications.

Includes index.
1. Data transmission systems. 2. Computer networks.
I. Title
TK5105.L43 1982 621.38′0413 82–47683
ISBN 0–669–05555–7

Published simultaneously in Canada

Printed in the United States of America

International Standard Book Number: 0–669–05555–7

Library of Congress Catalog Card Number: 82–47683

For my parents, Raymond, Linda,
Flora, Tom, and their fellow
visionaries of the twenty-first century

Contents

Figures and Tables

Figures

Tables

Preface and Acknowledgments

With increasing frequency, the news media and trade magazines report the availability of new products and services that have the potential to alter our traditional life-styles. The variety of new consumer electronic products has been made possible by technical achievements in the laboratory and the ambitious efforts of entrepreneurs. The public has responded by buying personal computers, two-way cable-television services, video recorders, and word-processing machines in growing numbers. The prospects appear to be equally bright for new mobile telephone services, direct-broadcast satellite systems, videotex, and electronic funds-transfer services. These developments will allow consumers to shop and bank electronically, access large information databases and electronic libraries, and receive home-entertainment services. The appearance of these products and services in the home moves us ever closer to an electronic-cottage society.

Modern societies have been shaped by important infrastructural developments, and this electronic society will be no different. This book examines some implications of innovative message-transfer technologies. Chapter 1 presents message-transfer developments in the context of an evolving information society. Chapters 2 and 3 describe the technology, traffic, and use of such services. Chapter 4 analyzes market-organization issues, emphasizing the question of service provided by the U.S. Postal Service. Chapters 5 and 6 examine the impact on postal and office activities. Chapter 7 focuses on the liability and privacy implications of electronic message services. The final chapter presents a summary evaluation of electronic message-transfer developments.

The electronic message-transfer environment is changing extremely rapidly. New services and equipment with more capabilities for lower prices continue to appear. The discussion in this book focuses principally on the emerging trends in message transfer and their implications rather than on a detailed description of each new service. The trends clearly suggest that such services are becoming an important part of our evolving electronic society.

It is difficult to acknowledge all of the people who have made important contributions to this book. The late Raymond Bowers inspired my pursuit of many of the ideas developed here. I am especially grateful to Arnim Meyburg, whose supreme efforts helped make this book possible. Special thanks are due to Philip Bereano, Richard Schuler, G. Patrick Johnson, Sheila Jasanoff, John Langley, and Lawrence Williams for their careful reading and constructive comments on earlier drafts of the manuscript. I am indebted to Ramesh Vaidya, Sara Edmondson, Wayne Lieb, and Steve Strong for their valuable research assistance, and to the numerous people who provided important insights and information. Finally, many thanks must go to Deborah Van Galder and Patricia Apgar for their efforts in producing this manuscript.

1 Introduction

The provision of effective and efficient message-transfer services has always been a vital national concern. Mail and telecommunications, together with transportation alternatives, represent critical components of the infrastructure of the United States. They facilitate social and economic interactions among the general public. Proposals for changes in the nature or scope of these services are frequently of national interest and often the subject of controversy.

Advances in electronics over the last several decades have led to the development of improved, efficient, and versatile communications systems that will affect message delivery in several ways. The telephone already provides some substitution for written message transfer, but this is a modest portent of the future. Modern electronics and recent technological developments in information processing and telecommunications will change conventional mail operations. Furthermore, emerging electronic message-transfer systems will offer radically new message-service options. Finally, electronic technology is altering and blurring the traditional definitions of message transfer.

Interest in electronic message-service capabilities has been developing in various sectors. An increasing number of large, geographically dispersed organizations already use electronic message-transfer technology in private, transcontinental, and even worldwide telecommunications networks. These intraorganizational activities suggest that electronic techniques provide economic and operational advantages over conventional ones. The success of intracompany electronic-transfer systems has prompted various private vendors to offer new public electronic-transfer services. Although these undertakings are intended to reach primarily the business community, one can easily envision the expansion of services to a more general public within a decade.

Indeed, various telecommunication carriers are rapidly expanding their services to reach a wider audience. Cable-television distribution companies are designing two-way systems that can be adapted for message transfer needs. Developers of videotex systems are planning services that can utilize two household fixtures—the television screen and the telephone—for electronic message-transfer applications. It is not surprising that market-research firms project a growth in annual industry revenues from the current level of $70 million to $2 billion by the end of the 1980s.

Postal authorities have maintained an active interest in electronic message-transfer possibilities for over two decades. Since the Speed Mail facsimile experiments of the late 1950s, the United States Postal Service (USPS) has cooperated with Western Union in operating two electronic message services, Mailgram (introduced in 1970) and ECOM (introduced in 1982). Longer range planning efforts, beginning in the 1960s, have continued the development of a more sophisticated service concept. USPS has already spent about $19 million to develop a system that is expected to cost $1.8 billion by the mid-1990s. Postal officials are clearly interested in the prospects for electronic services.

The development of electronic message-transfer systems by both the public and private sectors offers many possibilities. An increased use of electronic services will affect USPS organizational requirements regardless of whether it is permitted to continue electronic-service development. Electronic-transfer systems offer the potential to conserve scarce national resources. There may be equally important legal consequences. For example, liabilities and responsibilities will have to be assigned for harms resulting from the use or abuse of the new technology. A potential exists for new harms such as electronic message eavesdropping. There are also jurisdictional questions regarding the regulation and control of electronic message transfer.

The impacts of electronic message transfer will extend far beyond the postal and telecommunication industries to those involved in the production, handling, and transfer of information and messages. For example, office workers, who comprise about seven percent of the national workforce, may face restructured job responsibilities and changing workplace conditions. The prospect of home terminals becoming mass consumer items also suggests that electronic message-transfer developments could affect the nature of human communication activities, patterns of information distribution, and public access to computer processing.

Communication via electronic message-transfer systems may raise a wide range of policy questions, such as the desirable direction of social evolution, the structure and distribution of costs of large technological systems, and the nature of social progress. Emerging electronic systems may alter expectations and perceptions about the nature of the message-transfer process and about message-transfer speed requirements. Users may expect more rapid replies to correspondence. Electronic message-transfer developments could accelerate the emergence of the wired electronic society, promoting suggestions that the communicating public may gradually lose subtlety and nuance in message exchanges.

The use of new communications technology may reduce the cost of written message-transfer services and increase the speed of message communications. However, such technologies should not be adopted solely on the basis of cost reductions or speed advantages. The effects of technological change often reach far beyond the primary areas of intended application, which, in turn, may impede the subsequent implementation of socially efficient technologies. Because written message-transfer services are widely used in the conduct of daily activities, the appearance of a radically new alternative to conventional transfer services suggests that a broad range of effects is possible. Engineers, planners, and other decision makers need to understand the indirect, unanticipated, and delayed consequences of new technological alternatives to design and implement such services effectively and in a socially beneficial manner. By examining these consequences while the technology is still under development, contingency plans may be made that address the problematic issues before they become the focus of social conflict and political debate. This exploratory assessment considers the implications of evolving electronic message-transfer services, both its costs and benefits.

One might expect an assessment of electronic message transfer impacts to be limited to effects resulting from the transfer process. However, an impact analysis that separates the transfer activity from the production and use of text or messages and the processing of information is artificial. The conversion of text or messages to and from machine-readable code necessarily involves information producers and consumers. Also, the use of specialized equipment *solely* for word processing (that is, text production or manipulation), accessing a computer, or message transfer is rarely justifiable because all three functions can be performed together, with only a small increase in costs, causing them to be considered virtually inseparable.

Their proponents suggest that such services will increase labor productivity of message handling in the transmission stage as well as in the production, storage, and retrieval of messages. In addition, they anticipate that the new media will permit a conservation of scarce resources, long-run decreasing relative costs to users, an overall improvement in business efficiency, and closer interaction among individuals and organizations. Less optimistic observers caution that new electronic message services could further dehumanize the workplace of information workers, create new situations of liability, threaten individual privacy and freedom, and increase the gap between the rich and the poor. The discussions in the following chapters consider these important possibilities.

2 Electronic Message-Transfer Technology

The recent emergence of innovative electronic message services has been made possible largely by a confluence of technological developments in several fields, including telecommunications, information processing, and solid-state technology. As a result, a wide variety of ideas exist concerning the services and markets to be developed. Much of the underlying confusion is due to the fact that various electronic service concepts have originated from very diverse sources. Office managers see electronic developments as a potentially cost-efficient means of handling letter correspondence. Computer and telecommunication specialists envisage electronic services as evolutionary developments in message or data transmission. Terminal vendors and office-equipment suppliers project a large peripheral market. Finally, postal officials foresee a technology that can provide new services and also alleviate budgetary problems resulting from rising labor costs.

The majority of enthusiasts presume that electronic technology will provide *end-to-end* communication with input and output devices that can perform a variety of other functions besides transfers. For example, the introduction of microelectronics into office typewriters allows them to serve as electronic message terminals. With the appropriate hardware, these enhanced typewriters can be used for semi-automated document preparation and handling. In essence, electronic message systems might be regarded as integrated text production, text management, and message communication technologies.

Access to end-to-end systems minimally requires the use of rudimentary terminals. Casual users may be unable or unwilling to accept the relatively high, fixed costs of terminals plus the additional costs for telecommunications that would be necessary to use a system on a regular basis. As a result, the initial market for end-to-end electronic services may be somewhat limited. Large corporations with a sizable volume of communications flowing between facilities may install private systems. In addition, end–to–end electronic services may exist for specialized markets where speed or convenience of message transfer is highly valued (for example, interbusiness communications or funds-transfer services).

One form of electronic service is intended to serve a wider audience than a business audience. This hybrid configuration can be designed to provide transfer services for use by the general public without purchasing or leasing terminals. With this goal in mind, USPS, among others, is studying the feasibility of deploying a sophisticated nationwide hybrid system that uses a combination of electronic and conventional physical methods to transfer messages between sender and receiver. At present, USPS provides two rudimentary hybrid services, Mailgram and ECOM. If consumer reaction is favorable and political forces do not impede subsequent development, USPS may implement a sophisticated hybrid service similar to the one described later in this chapter. Of course, private firms may offer such services if consumer demand is sufficient.

End–to–end and hybrid electronic message-transfer systems will provide a wide variety of technological capabilities. Each type of system will use different hardware configurations. For instance, an integrated text-production and message-communication system might require more storage facilities and higher-quality printing equipment than a system that is solely communication-oriented. Similarly, paper-based hybrid electronic systems may require more paper-handling equipment than end–to–end electronic systems. Hence, it is necessary to clarify a definition of electronic message services for the purposes of this study.

Electronic message transfer services transmit person–to–person communications that are graphic- or alphanumeric-character-oriented and digitally encoded. Such services convey messages electronically but at some stage may produce and physically transport a paper copy by conventional means. Computerized data-retrieval systems or computer-aided instruction systems are not considered electronic message-transfer systems in the context of this study because such systems do not communicate between humans.

Several rudimentary electronic services are already in operation. The oldest hybrid service is undoubtedly the telegraph service, which began to evolve rapidly during the mid-nineteenth century in this country. Probably the fastest-growing hybrid electronic service is the Mailgram, which is jointly provided by Western Union and USPS. Facsimile machines connected by the telephone network are end-to-end transfer systems. Other familiar examples include Western Union's Telex and TWX services. Although all these services adequately meet our definition of an electronic message service, table 2–1 reveals individual differences in the quality and features of various available services. Technological trends indicate that new electronic systems under development will offer faster, much cheaper, and higher quality services than their predecessors. The following discussion will concentrate principally upon these emerging systems.

Table 2–1
Selected Features of Antecedent Message-Transfer Services

	1979 Message Volume (000)[a]	*Transmission cost per message*[b]	*Speed*	*Access Requirements*	*Comments*
Telex TWX	1470	$3.50 for 600 characters.	Immediate or short delay	Terminal installed on users' premises	Telex & TWX services are operated by Western Union. TWX lines are three times faster than Telex.
Facsimile (FAX)		$1.50–$2.00 per page.	Immediate	Terminals installed on users' premises	Page transfer costs dependent upon complexity of FAX units.
Mailgram	33,465	$3.90 for 50 words or less. $1.25 more for increments of 50 words.	Overnight	Terminals located at Western Union facilities accessible by telephone. Conventional postal delivery.	Operational since 1970. Uses Western Union acceptance and transmission with USPS postal delivery.

[a]Data not available for FAX.
[b]Excludes terminal costs.

The Emergence of Improved Electronic Message-Transfer Systems

Several technological factors support the evolution of improved electronic message-transfer systems. Advances in solid-state technology have led to the development of small economical and reliable microprocessors that now have a computing power equal to equipment developed in the 1950s at a cost of several million dollars. Over the past several decades, computation costs have been declining at a rate of 60 percent annually. As a result, microprocessors are now being used to distribute computing power throughout electronic message-transfer systems. This change in system architecture has nurtured the development of terminals with increased sophistication and added capabilities. The use of microprocessors in electronic systems also allows messages to be processed prior to transmission, which implies the more efficient use of communication channels. These applications and other microelectronic developments have resulted in an increase of technical capabilities by 14 percent annually, while annual costs have fallen 12 percent.[1]

A second technological factor is the development of new communication techniques. It is now becoming economical to send messages via satellite and through glass fibers, both relatively new transmission modes. Even on the current prevailing communications media, new techniques have made it possible to transfer more messages per channel, with fewer errors and at a more rapid rate than previously attainable. These and other trends have made improved transfer systems both technically feasible and economically viable.

The Technology of Electronic Message Services

A recent survey of electronic message systems identified the functional characteristics associated with every system.[2] This survey suggested that every system performs information input, information output and display, and transmission. In addition, such systems may provide switching, text editing and manipulation, message filing and retrieval, privacy protection, special addressing and distribution, and computer coordination of information, flow, and content processing.[3] In this section, the operations of the hardware and software associated with these functional characteristics will be described.

General System Components

The technology utilized in any electronic message transfer system can be generally categorized according to four subsystems:

Input subsystem—an assembly of equipment that permits users to enter messages to be conveyed into the system so that they can be readied for transmission

Transmission subsystem—equipment that will transfer messages economically, accurately, and dependably

Output subsystem—a package of devices that allows users to receive messages in intelligible form

Control subsystem—hardware and software that will both store, prepare, and trace messages entering the system, and coordinate operations of the total system, including input, switching, signal processing, transmission, and output.

Input Subsystem. Messages that enter an electronic message system must be translated into some form of machine-readable code. Various input devices can accomplish this encoding process. Modified office typewriters, word processors, computer terminals, and other keyboard equipment with either hard-copy or video display can serve as convenient input alternatives with the addition of relatively simple hardware modifications. By adding a communication option and buffers when needed, these character-encoding devices prepare messages for subsequent transmission. When a message draft is typed on this equipment, each keystroke generates corresponding character bit codes (for example, the seven-bit ASCII codes), which convert message contents to binary stream data.

The relatively widespread deployment and low cost of keyboard devices makes them particularly attractive as input equipment for electronic message services. In the future, as costs of microelectronic elements continue to fall, keyboard devices with integrated microprocessors and memory will readily serve as electronic message-input terminals and also accomplish other office or household functions.

Messages composed of alphanumeric characters that have not been encoded during typing may be character-encoded by using special equipment. These translation devices, called optical character-recognition (OCR) units, contain microprocessor devices that compare the similarities of each read symbol against a stored master set of character possibilities. The microprocessor looks, in particular, for agreement in geometry of character strokes or matching tonal qualities of symbols (such as the congruence of light and shaded areas). If the similarities are close enough, the microprocessor decides the identity of the read symbol and the corresponding character code string is generated by the OCR unit.

The accuracy of OCR units depends on several conditions. Because all OCR techniques seek to match input symbols with a master key set, the input possibilities must correspond to those defined by the master set (for example, Greek letters cannot be identified and subsequently encoded if they are not included in the master set). In addition, print and paper quality must be chosen so that characters are easily discernible because the print contrast of input affects OCR readability. Furthermore, characters must by typed with a restricted set of typefaces or fonts. By limiting the stylistic qualities of the input characters, OCR units can encode messages more rapidly and with many fewer errors.

If these conditions are met, OCR encoding can be accurate. Yet, even under optimal conditions, errors are unavoidable. One study of OCR equipment performance concluded that under good conditions 2 percent of all characters read would be uninterpretable while 1 percent would have wrong characters identified and encoded. Under unfavorable conditions, the percentages were 10 percent and 2 percent, respectively.[4] At present, the greater cost and comparatively high error rate of OCR units limit their use. Although both OCR costs and error rates are expected to fall, character encoding of messages during the typing stage will continue to prevail.

Messages that are not amenable to character encoding can be image-encoded by facsimile equipment. These units accept messages that are not alphanumeric and that do not meet the preconditions for OCR readability. Facsimile units are also used to transfer and exactly reproduce formatted messages.

The facsimile technique was conceived during the 1840s. Since that time, units have been developed which use various methods to scan and image encode a page. Basically, these processes move either the input page or the scanning spot, or both, so that all the pixels on a page can be read and encoded in an order sequence.[5] A variety of commercially available facsimile transmitters can do much more than simply encode images in black and white. Some machines can handle halftone shading; others can process color images. Color-facsimile machines are much more expensive and have larger page-bit requirements than simple one-color equipment.

Facsimile transmitters have been classified in terms of the speed in which each 8½ × 11 page can be image-encoded and transferred over standard voice-band channels. Basically, these machines fall into three categories: 4 to 6 minute, 2 to 3 minute, and 1 minute or less. The principal difference among these devices is not the scanning rate, however. Differences in transmission speed are due to the variety of data-compression techniques used in the encoding process. Although a sophisticated facsimile machine with integrated data-compression hardware allows more messages to be transferred over a given channel than uncompressed equipment, it also tends to be more complex and more expensive. Yet the added costs of compression will often be more than balanced by the savings in transmission costs.

Several conditions determine the quality of facsimile-encoded input. First, sufficient contrast must exist between the page contents and the page background so that the scanning device can differentiate among tonal qualities. Second, pixel size must be small enough to guarantee that the desired image resolution is produced. In other words, each pixel can be no larger than the width of the finest line occurring in a message. But as pixel size decreases, for a given page, the number of pixels necessary to cover that page must increase, thereby increasing the bit requirements of each page. As a result, some compromise must be made between resolution quality and page-bit requirements.

Facsimile units are very convenient input devices to use for encoding messages, perhaps suggesting a wide sales potential. However, because relatively large message page-bit requirements generate high page-transmission costs, the growth of deployed facsimile units has not been as rapid as expected. Trends indicate that faster machines that use more sophisticated compression techniques will continue to be developed. Projections discussed in chapter 3 indicate that costs of all types of facsimile machines will continue to fall and that volume of units in use will rise rapidly. However, all units may not be able to interconnect with each other because common standards do not currently exist for all machines.

Input subsystems contain more than message-encoding and compression equipment. A total input subsystem includes microcircuits that insert error-checking bits into message-bit streams so that transmission errors can be later detected. Various memory devices, such as buffers or memory registers, are also necessary for smooth system operation. Memory registers hold transmitter status information that must be passed to the control subsystem during message transmission. Such information includes timing signals (that is, synchronization characters or stop-start signals), transmitter sending speed, character-encoding bit length, applicable parity rules, and other supervisory data. Buffers are necessary to hold encoded messages (and sometimes packet-addressing and sequencing information) before transmission and during the error-checking process.

Other equipment is commonly incorporated. In situations in which input facilities utilize local telephone access lines, modems must be included in subsystem assemblies. If the message system employs sophisticated digital techniques such as packet switching, a microprocessor unit must divide encoded message data into packets and label each packet with address and sequence information. Level converters are often included in input subsystems to reduce the voltage and current of signals received from the control subsystem to a degree acceptable by terminal equipment. Furthermore, because declining cost of microelectronics has allowed processing to be distributed throughout a message system, input subsystems now include microprocessors which contribute to the supervisory and control functions of the control subsystem.

Finally, input subsystems may contain hardware and software that increase message security. A relatively secure product-ciphering technique, called Data Encryption Standard (DES), has been recently developed which uses a key length of 56 bits. A micro chip has been designed to implement the DES algorithm at a relatively low cost. Some have argued that this single key technique is inferior to the public-key systems that are now being proposed.[6]

In the future, public key ciphering that uses both encrypting and decrypting keys to augment message security may become standard implementations in input subsystems. Briefly, all encoding keys can be publicly published while decoding keys remain secret. To transmit a message, the sender encodes with the recipient's encoding key. Then the message can be read only by the recipient with the secret decoding key. One byproduct of this cipher system is that electronic signatures can be transmitted by encoding a message with the sender's decoding key prior to encoding with the recipient's decoding key. The recipient can then verify the sender's identity by decoding the message with the recipient's secret key and decoding it with the sender's public key.[7] A product-ciphering technique with a larger key length may be developed and implemented if the current DES algorithm proves to be insufficient.

Transmission Subsystem

Various transmission alternatives are available to convey encoded messages. The more rudimentary operations use the public switched-telephone network or WATS lines to connect input facilities with subsystems on an ad hoc basis. In such situations, users with compatible terminal equipment connect on a need basis and incur only standard toll charges that are determined by time and distance. The channel capacity of any connection established over the ubiquitous public switched-telephone network can be as high as 4,800 bits per second.

Users requiring transmission of a higher volume of messages among a limited set of locations can meet their needs by leasing private lines that connect each pair of locations. Private lines can have several advantages over switched connections. First, if a private line carries sufficient traffic, it can be more economical than the switched network. More important, the electrical characteristics of these analog lines can be adjusted to achieve an optimum data-transfer rate. Without such "conditioning," line distortion can attenuate different frequencies and vary the propagation speeds of arriving frequencies. Conditioning equalizes these differences, allowing channel rates of up to 9,600 bits per second.[8]

It is possible to increase the flexibility of private lines with the addition of specialized equipment. For instance, a private line service could be established between Chicago and Los Angeles that can be accessed by local area users over the public switched network. In this case, control equipment at the transmission location could queue and combine messages so that line capacity is fully utilized. (The operation of this control equipment will be described further in our discussion of the control subsystem.) This "private-switching" alternative can also be used with other private channels, such as microwave radio links and large-capacity digital channels.

In an effort to offer higher capacity, low-error transmission alternatives, carriers have developed wide-band digital channels. These connections, which may be either switched or unswitched, can sustain data rates of up to 230,400 bits per second and require no modems incorporated in the input or output subsystems if directly used. However, in situations in which digital channels are used for long-haul transmission only, with the public switched network serving as an access medium, modems are required at long-haul transmission and reception facilities as well as at input and output terminals.

Traditionally, wideband digital channels have been either copper wire, coaxial, or microwave radio link. For instance, Type 5700 leased lines are available with a capacity of 230,400 bits per second; the Type 5800 lines have an even higher potential rate. Both AT&T and Western Union offer switched wide-band services in a limited number of cities. These lines are capable of transmitting 56,000 bits per second and 38,400 bits per second, respectively. It is likely that such switched wide-band lines will eventually be able to support 100,000 bits per second at low error rates.[9]

A second high-capacity alternative involves using a microwave radio link. Such subsystems have provided reliable and economical channels that can support many million bits per second of traffic. Microwave communications systems rely on "line-of-sight" transmission and require repeaters mounted on large towers spaced between 10 and 100 miles apart, depending on terrain and weather conditions. These repeaters accept transmissions, regenerate the received message, and relay data to the next repeater until the intended receiver is reached. Microwave data communication services are being offered by a number of specialized common carriers and require electromagnetic spectrum allocations to avoid interference with other radio equipment.

Satellite communication subsystems represent another wide-band radio link for conveying encoded messages over a high-capacity system.

Basically, a satellite communication subsystem consists of one or more satellites moving in synchronous orbit (that is, moving at the same speed as the earth, allowing the satellite to seem stationary with respect to the earth) 22,300 miles in space, and two or more transceiving earth stations. Each earth station transmits messages to a satellite on one frequency and receives messages relayed by the satellite on another. It is possible to cover one-third of the earth with one satellite having a sufficient beamwidth. However, satellites with more focused beams (that is, a narrower beamwidth) cover less area and generate a higher ground signal at the same transmission power, allowing receiving stations to use smaller antennae than in the less focused configurations. In addition, this focusing technique permits the possibility of frequency reuse over large geographic areas, an important consideration as electromagnetic spectrum becomes an increasingly scarce resource.[10]

The principal advantage of the satellite alternative is that the costs of wide-band service are almost independent of distance. Studies of future costs have shown that messages traveling over ground distances greater than 300 miles will have costs which are insensitive to distance. Message costs will depend largely upon the number and costs of earth stations.[11] As the costs of earth stations continue to fall, satellite transmission will be an increasingly attractive means to convey messages electronically. In fact, a recent study that compared satellite transmission to terrestrial alternatives concluded that the former was cheaper except in cases of very low transfer volumes.[12]

At present, various domestic satellite systems are available for message transfers. Several others are under development. The general characteristics of current or planned systems are summarized in table 2–2. Each system has numerous transponders available, offering a very large amount of bandwidth per satellite. The more advanced satellites will have much greater capacity than their predecessors.

One other rapidly developing transfer technology appears to be promising for wideband transmission. Several carriers are developing optical fiber networks. When coupled to a specialized light source (such as a laser or light-emitting diode) that is modulated by message signals, these silicon-based links can transfer data at a rate of over 100 million bits per second. This technology is advancing so rapidly that data rates are expected to increase by five times while fiber costs per meter are projected to fall by a factor of ten by the end of the decade. Thus, optical fiber transmission may begin to replace "wired" alternatives.[13] As optical fiber transmission becomes cost competitive with other alternatives, electronic message systems will probably incorporate such technology.

Various transmission media are available for message transfer. Individual transfer systems may rely on the public switched network or

Table 2–2
Domestic Communications Satellite Systems

Carrier	*Western Union*[a]	*RCA*	*COMSAT*[b] *AT&T/GTE*	*Satellite Business Systems*	*Southern Pacific*
Number of planned orbiting satellites	3/2	5	6	3	4
Transponders/satellite	12/18	24	24	10	24
Total useful bandwidth/satellite (MHz)	432/1782	816	816/864	430	1296
Frequency band	C/K	C	C/K	K	C/K
Earth station antenna size (meters)	15.5	4.5–10	30	5.5–7.7	

[a]Western Union proposes to operate both WESTAR and ADVANCED WESTAR satellites.

[b]AT&T and GTE now use COMSTAR satellites owned by COMSAT General. In the future they will use their own satellites.

private lines, depending upon specific needs. In higher volume situations, digital satellite, microwave, wire, and even fiber optical channels will be available in the future. These high-capacity systems may utilize the public switched network to increase local access to facilities. Such systems may use both synchronization techniques to increase channel capacity and message-switching to increase channel utilization efficiency. Of course, any of these specialized techniques requires the introduction of more processing power in any system and more control subsystem equipment. Table 2–3 summarizes message-handling capabilities of various alternative transmission channel links.

Output Subsystem. The output subsystem has several principal functions, including accurate reception of encoded transmissions, regeneration and assembly of messages, and video or hard-copy display of received messages. Much of the output equipment necessary to fulfill these functions will be housed together with input equipment in two-way terminal devices. In high-volume systems, however, terminals may contain soft-copy or screen display devices with separate supporting equipment designed for high-speed hard-copy message reproduction. To clarify these differences, the output subsystem is discussed separately from terminal equipment.

The output subsystem must coordinate activities with the control subsystem so that encoded transmissions are accurately received. Modems are necessary to demodulate signals if local telephone or leased lines are used. Timing signals (asynchronous or synchronous) must be passed by the control subsystem to the receiving equipment along with information about encoding details (that is, character or image), parity rules, packet sequencing (if used), and message security. Receiving equipment must confirm the accuracy of these control signals along with encoded message signals and then indicate to the input subsystem whether to continue to send new data or to retransmit garbled data. Microprocessors can be installed in the output subsystem to ensure that the reception equipment operates smoothly.

Message data that have been separated from control signals can be temporarily stored in buffers until they can be decompressed, decoded, and displayed by output devices. At this stage, the recipient can decide between temporarily displaying a message on a screen or producing a paper copy of the message. At present, video display units which are functionally similar to television screens are the dominant technology. In the future, these cathode-ray devices may share the market with various types of solid-state flat-panel screens. Some of these possibilities include plasma panels, light-emitting diode arrays, and liquid crystal devices.[14]

An output subsystem may contain equipment for hard-copy repro-

Table 2–3
Message-Transfer Capabilities of Alternative Transmission Channel Links

Link	*Link Capacity (1,000 bits per sec)*	*FAX pages/hr (400 k-bits/pg)*[a]	*Character encoded pg/hr (8 bits/ch and 2,400 ch/pg)*[a]	*Remarks*
Analog private line	9.6	86	1,800	Assumes line conditioning
Public switched lines	4.8	43	900	Can be used for entire send or receive connection or as access to long haul channel
Satellite channel	45,000	405,000	8,437,000	Weather conditions will occasionally lead to short transmission interruptions
Microwave channel	10,000	90,000	1,975,000	Weather conditions will occasionally lead to short transmission interruptions
Private digital link	230,000	2,070,000	43,125,000	A very high-capacity channel; other links offer 50,000 bits/sec and 38,000 bits/sec
Switched digital link	56,000	504,000	10,500,000	Available in limited number of cities only
Fiber optic link	500,000	4,500,000	93,750,000	Estimates

[a]These estimates indicate only order of magnitude because some link capacity must be devoted to transfer of supervisory and control signals.

duction. Two basic categories of printers produce paper copies: impact printers, which consist of a mechanical interface that strikes ribbon ink to paper; and nonimpact printers, which relies upon electrostatic, optical, or other noncontact techniques. Although traditional impact printers, such as those used in standard teletype or typewriter devices, are the oldest and most well-established techniques, they tend to be slower and noisier than nonimpact techniques. Facsimile recorders commonly use an electrostatic or electrosensitive printing method.

Future printing devices will use a variety of techniques. In high-volume situations, high-speed dedicated printers (those producing more than 1,000 lines per minute) will use electrostatic and ink-jet methods. Medium-speed printers (200–1,000 lines per minute) will continue to use impact techniques, while low-speed devices (less than 200 lines per minute) will employ thermal processes for message reproduction.[15] However, modified electric typewriters that employ conventional impact techniques to produce about 10 lines per minute will continue to be popular hard-copy alternatives in an output subsystem, especially when the volume of messages is low.

Control Subsystem. The control subsystem coordinates all of the activities of the other three systems and conveys encoded messages accurately and efficiently. The amount of processing and storage equipment required depends on the degree of sophistication designed into each system and also on service features. For instance, a simple facsimile system using several 6-minute facsimile units and the public switched network will require less control equipment than a high-volume satellite-based, public message-transfer system that conveys numerous messages to many points. This discussion will focus upon general categories of processing that this equipment must accomplish.

One other caveat should be made. The advent of low-cost microprocessor technology has allowed designers to place "intelligence" (that is, processor or computing power) at virtually any point in a system. As a result, total computing power need not be entirely contained in one centralized unit, but may be distributed to the input and output subsystems. The subsequent discussion emphasizes the functional rather than the locational nature of the control subsystem.

The emergence of high-capacity transmission channels has allowed the development of message-transfer services with large message volumes per channel. These wide-band channels also can transfer bits at lower cost than can narrower channels. As a result, many services are designed to utilize these more economical channels. But because few individual terminals have the message volume, transmission speed, or reception speed to utilize these channels fully, techniques have been developed to promote channel sharing. One of the chief functions of the control subsystem is to ensure that channels are used in an efficient cost-effective manner.

One way to achieve efficient channel sharing by many local terminals over a synchronous long-haul channel is to incorporate a time division multiplexer in the control subsystem. This equipment essentially apportions the use of a wide-band transmission channel to sharing terminals sequentially. For instance, if five sharing terminals are each given 1 minute of transmission time in every cycle, then after every 5 minutes of waiting, each terminal will be able to send or receive accumulated data during the next 1-minute slot. Time division multiplexers ensure that channel access is properly rotated, transmissions can be identified according to position in each cycle, and message continuity is maintained between senders and receivers because transmissions are not continuous.

Statistical multiplexing is a dynamic allocation method that relies upon the fact that terminal transmissions are usually brief, concentrated bursts and tend to use less than 10 percent of the connection time. Statistical multiplexers apportion time slots to active users and eliminate routine allocations to terminals not demanding channel time. Studies indicate that two to four times as many users can be accommodated with statistical multiplexing than with time division multiplexing. However, more storage and sophisticated control equipment is required.[16]

Concentration, a third technique, allows many terminals to share several channels. Essentially, concentrators poll each input terminal to find those which are awaiting channels for transmission or, correspondingly, inform output terminals that messages are to be received. When the terminals acknowledge these control signals, connections to the trunking channels are then initiated. More sophisticated terminal equipment holds entries in temporary storage devices before selecting trunk channels to ensure that a high channel load is maintained at all times. Concentrators use either circuit switching or the store-and-forward message-switching techniques mentioned in appendix 2A.

The control subsystem accomplishes various other functions besides promoting channel efficiency. Some message-transfer systems have equipment that allows dissimilar terminals to communicate. These systems have processors which accept and relay transmissions between terminals with different speeds. In addition, they can convert character or format (such as sequence of codes or code blocks) codes so that, for example, a 5-bit code machine can communicate with a 7-bit code machine. Conversion capabilities add flexibility to any system, but again require additional control equipment and buffering.

Another control subsystem responsibility is the sorting, routing, and tracing of messages after entry. Control equipment must ensure that appropriate switching functions (either message or circuit) sort messages, route them in the proper direction, and eventually deliver them to the required destination. Some have preferential service features that queue messages according to priority specifications. Finally, this equipment must control inventory and trace message movements to detect losses or erroneous transfers.

The central task of the control subsystem is the supervision and coordination of the system operations going on in the other subsystems. These operations include error control, encryption key assignments, synchronization and timing, and message-routing management. The control subsystem reacts to control signals sent by the other subsystems and ensures that the transfer system works smoothly. Table 2–4 lists some of the functions of a relatively sophisticated transfer system.

The fundamental components of the control subsystem include software, microprocessor devices, and storage elements. Software development and maintenance for message transfer applications can be expensive. However, existing software packages may be adopted economically. The rapid development and use of microprocessor intelligence will also reduce costs.

Electronic message transfer systems will need three types of storage devices: quickly accessible memory, which stores software instructions; mass storage to hold messages awaiting action (such as printing and transmission); and terminal buffers, which hold data until blocks are assembled and transmitted and which assist in error control. The amount of memory required for each system will depend on the sophistication of its design, the capacity of transmission channels, and the speed at which channel-sharing equipment can give terminals access to the system. For instance, satellite systems require several seconds to transmit signals between Earth stations. The use of wide-band channels require buffers

Table 2–4
Functions of the Control Subsystem

Functions
Multiplex/concentration
Switching and network control (packet or circuit)
Format, code, or speed conversion
Buffering
Routing, priority, and inventory control
Error control
Timing and synchronization
Message management (that is, segmentation and reassembly of messages)
Security key management
Link control
Terminal control
System diagnosis

that can hold a large amount of data during error detection. Memory technology, such as magnetic bubbles and charge-coupled devices, as well as new semiconductor production technology, will further reduce memory bit costs, although quickly accessible memory will continue to cost more per bit than mass storage devices.

Improved Electronic Message-Transfer Networks under Development

Various corporate entities are using the previously described technology to develop new systems for intracorporate or public use. Several of the larger end-to-end networks are described in this section.[17]

TYMNET'S Ontyme Service

The TYMNET system, which began operting in mid-1977, uses circuit switching to offer electronic mailbox service to more than 600 receiving nodes. Basically, messages are broken into packets and sent to a ''mailbox'' or file location in a centralized computer bank at speeds of up to 4,800 bits per second. The addressee must ask the computer for messages stored in his or her mailbox. Although this service is relatively inexpensive, message delivery to terminals is not automatic and requires explicit requests by the intended recipient.

Satellite Business Systems

The SBS joint venture undertaken by IBM, Comsat, and Aetna Life and Casualty offers wide-band switched capabilities that will mix facsimile with voice and high-speed data services in a digital format. The system will use satellite channels with small Earth stations (16- to 23-foot dishes) located on the customer's premises. SBS has already contracted with Hughes Aircraft to build three 14/12 GHz satellites. Because each Earth station will cost more than $300,000, SBS is being initially designed for use by multiplant corporations with large communications volume requirements.

Telenet's Telemail Service

The Telenet system is a large network that was recently acquired by GTE. This packet-switching carrier began operating in 1975 and expects to offer access to electronic-message and document-communications service over local telephone lines in 400 cities. Plans are being made to upgrade channel capacities from 56,000 bits per second to 1.5 million bits per second, and eventually to use packet satellite channels. Ideas similar to the SBS design are being developed.

AT&T's Advanced Information Systems/Net 1

AT&T will offer a data-communications network with electronic message-distribution capabilities through its American Bell subsidiary. This system will be designed with processing power capable of interconnecting five classes of terminals with varying transmission speeds, character code sets, and formats. Initially, the service is being developed for organizations with huge data communication requirements. AT&T's long-range plans are not yet clear.

Other Networks

Many other telecommunication-oriented firms, such as MCI, Southern Pacific Communications, BBN, Rapicom, Source Telecomputing, and Graphnet are planning to expand and improve their message-transfer facilities. Developers of new two-way cable television systems, personal computer, and videotex services are also designing their systems for electronic message-transfer applications. While it is difficult to project the full range of services that will be offered by carriers in the next decade, it is evident that there will be a proliferation of innovative terminal-to-terminal services.

Hybrid Electronic Message-Transfer Systems

All systems discussed so far provide all-electronic or end-to-end services. These system designs assume that terminal equipment is located in the user's home, office or factory. Messages are conveyed entirely in electronic form to recipients who can examine them on a visual display and decide whether to produce paper copies. Although these systems may save time, labor savings may be offset by terminal rental or purchase costs. These costs may be beyond the reach of the general public, especially when the volume of messages is low. As a result, several organizations, most notably USPS, have attempted to design innovative hybrid systems.

Hybrid message-transfer systems employ centralized terminal equipment so that costs can be shared among a large group of users. Messages are conveyed chiefly by electronic means. But since terminals are not installed on the user's premises, electronically encoded data received in a destination city must be printed on paper and physically transferred from receiving station to addressee. ECOM and Mailgram services are two rudimentary examples of hybrid services. The new hybrid system under study promises to be much more sophisticated.

The Postal Service has spent almost $19 million in developing an electronic message-transfer system design over two decades. The following discussion of the current design proposal describes how a fully deployed public hybrid system might be configured.[18]

Such a system could eventually provide a nationwide service with next-day delivery for 95 percent of messages entering the system by 5 p.m. on the previous day. Plans are also being made to provide a priority service (transfer of messages in 1 or 2 hours) at a premium price. Both the overnight and priority service would use the conventional postal system along with new electronic equipment under development.

Various input alternatives would be available. Large-volume users, such as banks or credit-card companies, whose messages mostly consist of computer-generated material, could drop off magnetic cards or tape containing encoded messages. USPS would provide equipment to read and route these messages electronically. Other heavy users with messages already printed on paper would submit paper packs (that is, cardboard boxes of message pages without envelopes) to be facsimile-encoded by USPS personnel. A user with a single hard-copy message could access the system by using a coin-operated facsimile machine or electronic mailbox that might be located in a post office or other public place (such as a shopping mall). USPS would accept messages submitted via local telecommunication links, although use of this option would not be encouraged.

Previous designs considered the development of OCR inputs. However, initial plans call for a limited use of OCR encoding (for address encoding of bulk paper messages only) because of the undesirable error rate. If future OCR error rates can be reduced, this input option may be more attractive.

Specialized terminal equipment is being developed to meet the system's unique requirements. A state-of-the-art facsimile machine now being tested will transmit paper-pack messages at a rate of 10 pages per second. Electronic mailboxes are being designed to offer 10 to 1 compression with a high resolution feature of 200 to 300 pixels per inch. Facsimile machines could offer up to eight-color reproduction, including several gray scales for black-and-white processing. In addition, paper-handling equipment is being developed to mechanize the movement of paper-pack messages through the encoding equipment.

Messages contained in paper packs or on magnetic media would be deposited at post offices or regional postal facilities (called sectional centers). Message would then be transferred by truck to specialized message centers for electronic routing to receiving centers. In the case of remote electronic mailboxes, encoded messages would be moved electronically to message centers over local lines or stored on magnetic media and trucked to centers, depending upon the cost-volume characteristics of each mailbox. The electronic centers would be geographically dispersed according to the volume of messages generated in each area. A recent study indicated that 87 U.S. stations would be adequate to deliver overnight messages. Each of these 87 interconnected stations is expected to cost $320,000.

Several studies of transmission alternatives have concluded that satellite communications are the most cost-effective linkages between electronic centers. Recent analysis has shown that leasing one satellite channel would be much less costly than using terrestrial channels. Each satellite channel using the 6/4 GHz band can handle 40 million bits per second. By dividing this wide-band channel into 80 subchannels of 500,000 bits per second each, modem and transmitter costs could be reduced and many simultaneous communications between pairs of centers could take place. It is estimated that one satellite channel would be able to handle 20 billion messages annually.

Messages arriving at destination centers might be conveyed to recipients by several means. Messages directed to user terminals would be routed to a communication processor and interface and held until contact is made between message centers and remote terminals. Messages to be conventionally delivered would be reproduced on paper, put in envelopes, and sorted for carrier delivery. After hard-copy production, messages would be trucked to the local postal distribution stations and fed into the conventional mailstream. Hard-copy message production would require the use of sophisticated paper-handling equipment to facilitate the movement of paper through the output subsystem.

The current design is estimated to require an investment cost of $1.8 billion and have an annual operating cost of $285 million. It could carry 25 billion messages annually at a transmission and processing cost of 1.8 cents per message. Delivery costs will be additional. If the carrier cost per piece is maintained at the current 7 to 8 cents, then messages could be conveyed over this hybrid system at a unit cost of less than 10 cents. A hybrid system design is illustrated in figure 2–1.

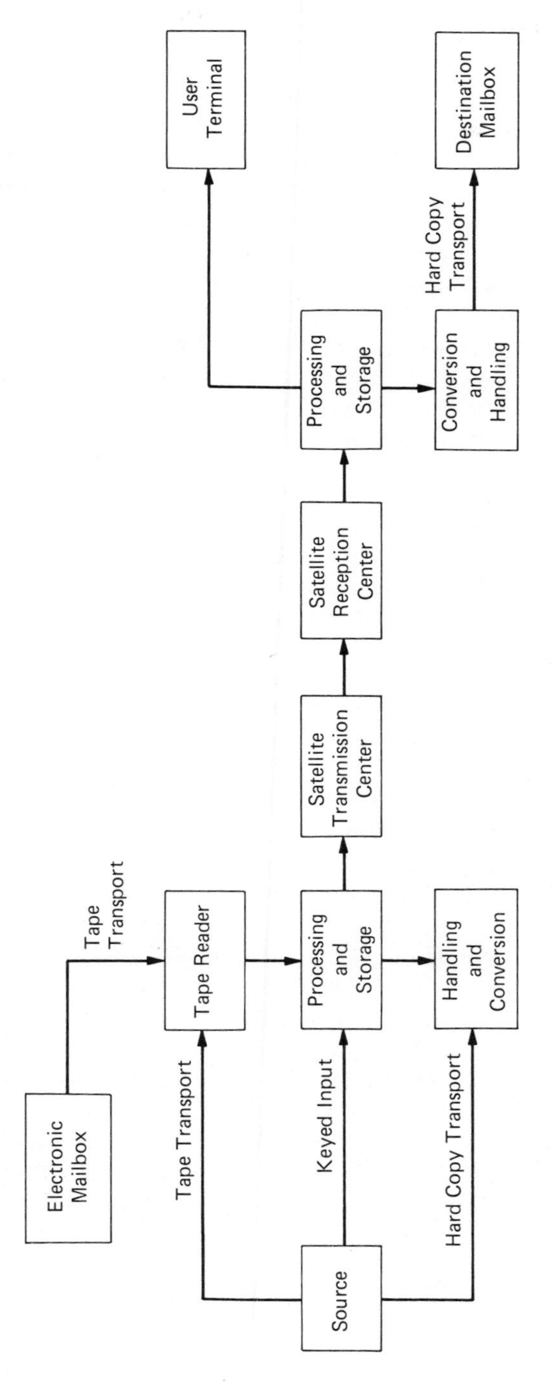

397 **Figure 2–1.** Electronic Hybrid System

Low-Cost Terminals for the General Public

Hybrid systems such as the design proposed by USPS could evolve into all-electronic end-to-end systems in the distant future. Many businesses could have remote terminals by 1985, permitting both interbusiness and intracompany message transfers.[19] An increasing number of firms already have data terminals and microprocessor-equipped typewriters that can be adapted for message transfer use. Trends indicate that acquisition costs are not prohibitive for business and thus, terminal purchase or lease will continue to increase as costs decline.

It has been suggested that low-cost terminals will eventually be developed for home use. But public acceptance of this technology will depend quite heavily on price as well as ease of use. In other words, users must be able to operate the technology with a minimal amount of training and services must be perceived as competitively priced with alternatives. If terminal and service costs drop below conventional mail costs, home terminals may become as ubiquitous as residential telephones.

A rudimentary home terminal consists of a display device containing a decoder, storage, a character generator, and control unit. This equipment could connect to a transfer network (such as the telephone network) by means of an interface device (such as a modem) and a touch-tone telephone for control signaling and generation of preset messages. Optional equipment could include a keyboard to make typing messages easier, a printer for producing paper copies, or portable memory to store messages electronically.

Several companies are currently developing microcircuitry that permits a standard television set (now owned by 97 percent of all households) to act as a rudimentary terminal with visual display. Industry research on designs of terminals for home information retrieval suggests that adaptor costs will drop below $100 by the middle of the 1980s. A terminal device with full alphanumeric keyboard, hard-copy printer, and tape-cassette storage could be available for under $500 and rent for $10 to $30 per month (depending on depreciation and maintenance arrangements) during this decade.[20]

The development of keyboard-oriented terminals assumes that their users can type. Because typing is not a universal skill, several companies are developing low-cost facsimile units for home use. The French postal-telephone administration has planned a 1-minute unit which will cost under $1,500. Several Japanese companies are developing home facsimile units for under $300. If these efforts are successful, most households may have keyboard and facsimile units.

Beyond the Transfer Function

A wide variety of related technologies deliver more specialized services than simple transfers of messages. These systems, such as word-processing, computer-teleprocessing, transaction-network, and information-retrieval systems, commonly contain much of the hardware and software necessary to implement a message-transfer service. Often a small amount of additional equipment and software will expand these systems into multipurpose facilities. The proliferation of these technologies could influence the development of electronic message-transfer services.

At the present, corporate management is attempting to increase office productivity and lower administrative costs. The introduction of word processors is directed toward this goal. Word processors are designed to allow text to be easily edited and manipulated, thus substantially reducing the amount of typing necessary to revise and produce new drafts of the material. There are four basic categories of word-processing equipment in current use:

Stand-alone hard copy—less sophisticated equipment for automatic typing of repetitive letters, merging paragraphs, limited editing functions, and removable storage.

Stand-alone display—similar to hard-copy stand-alones but with cathode ray displays (from one line to one page) and printers to produce hard copies.

Shared logic—minicomputer with rudimentary terminals or connecting stations.

Time-shared—data-processing services with word-processing options.

All these systems have input/display units, information storage, a control unit, and a hard-copy printer. The addition of low-cost (from $1,000 to $4,000) communications equipment to such systems creates integrated economical message preparation and communication terminals that can be interconnected by message-transfer channels.

A second technology that can influence the development of message transfer systems are the home and business information-retrieval systems, either videotext or teletext systems. Several companies are currently developing systems that use telephone lines and modified television sets similar to the Prestel or Antiope systems. GTE has obtained Prestel licensing rights for the United States and Canada. Knight-Ridder and AT&T will test-market a similar home information system designed to deliver news, sports results, weather, and calendar of events into up to 200 Miami homes. In addition, a home television/telephone service is being tested in Kentucky that will inform farmers about weather conditions and other crop-related information. All these experimental systems will have software and hardware that can be adapted for message transfer.

The proliferation of personal or remote computing is another factor that may foster the development of message-transfer services. The advent of teleprocessing created a need for and supported the operation of early message-transfer services. Since then, remote teleprocessing has grown to a point that low-cost time-sharing services are available for home use after business hours. One can expect teleprocessing services of the future to offer access to computer programs which, for example, provide entertainment (such as video games), maintain a record of financial transactions, and edit text, as well as transfer messages.

These technologies may have a very real influence on the nature and popularity of electronic message-transfer services. The average household receives only 150 pieces of first-class mail per year, hardly a sufficient volume to justify the purchase or lease of a terminal acquisition solely for transferring messages. However, consumers may have more incentive to acquire terminals that have multiple uses, including message transfer. Thus, access to all electronic transfer services may be conditioned by the proliferation of other terminal-related activities desired by consumers.

Technical Issues

It is clear that sufficient technical knowledge and necessary component technologies are available to implement electronic message transfer systems which meet the most rudimentary or very sophisticated requirements. Indeed, various public and private organizations are moving rapidly to build such systems. There are, however, various technical issues which require explicit attention.

The developing systems may utilize differing technical characteristics. For instance, message-encoding techniques, transmission speeds, and control signals may vary among different terminal devices, transmission alternatives, and control subsystems. If it is desirable to ensure that most hardware and software packages are compatible, equipment manufacturers and system designers must be encouraged to agree on certain standards. However, it may be premature to set standards during the early period of development. Alternatively, conversion equipment could be built into emerging networks to ensure compatibility.

New systems will undoubtedly have some physical means to protect the security of messages. Great sums of money can be devoted to develop hardware and software for message security. Alternatively, the government could enact new wiretap or privacy legislation to protect message transfers. This choice is basically between physical and social technologies.

If system operators decide to use key ciphering to secure messages, is the DES key sufficient for most public uses? DES hardware is already available, but some suggest that public key codes are necessary. Another related issue concerns key management. Who should control distribution of keys and who is responsible for errors if they occur?

Various firms propose to use radio links (satellite or microwave) to transfer messages. Studies showing that satellite channels are cheaper than terrestrial channels are based on allocating the electromagnetic spectrum without charge. This allocation method has functioned smoothly as long as supply has exceeded demand. Yet demand for spectrum use is increasing continuously. Should annual charges for spectrum use be levied as a market solution to this allocation problem? Should electronic message-transfer services be allowed to use radio frequency links at any price if wired alternatives are available? A comprehensive and rational allocation scheme will be increasingly necessary as spectrum becomes ever more scarce.

Technical issues such as these must be resolved along with more detailed implementation questions to develop innovative electronic message-transfer systems that are socially and technologically efficient. Subsequent chapters will examine some of the legal, social, economic, and political dimensions of electronic message-transfer development.

Notes

1. U.S. House of Representatives, *Postal Research and Development* (Washington, D.C.: U.S. GPO, 1978), p. 34

2. A general primer on technical aspects of electronic message communications can be found in appendix A for readers unfamiliar with basic concepts or terminology.

3. Kalba Bowen Associates, *Electronic Message Systems: The Technological Market and Regulatory Prospects,* submitted to the FCC (April 1978), p. v.

4. General Dynamics/Electronics, *Study of Electronic Handling of Mail, Equipment Survey* (Washington, D.C.: NTIS, 1970), pp. 3–14.

5. Ibid., pp. 3–4.

6. D.E. Denning and P.J. Denning, ''Data Security,'' *ACM Computing Surveys* vol. II, No. 3 (September 1979), p. 224.

7. Martin Hellman, ''The Mathematics of Public-Key Cryptography,'' *Scientific American* (August 1979), pp. 146–157. Also see Robert Coven, ''New Cryptography to Protect Computer Data,'' *Technology Review* (December 1977), pp. 6–7.

8. See J.E. McNamara, *Technical Aspects of Data Communication* (Maynard, Mass.: Digital Equipment Corporation, 1978), pp. 347–352, for discussion of the six conditioning types available along with achievable bit rates.

9. Philco-Ford Corporation, *Conversion Subsystems for the Electronic Mail Handling Program, Task I* (Washington, D.C.: NTIS, 1973), pp. 4–9.

10. For more discussion of frequency reuse see J. Frey and A. Lee, ''Technology of Land Mobile Communications,'' in *Communications for a Mobile Society—An Assessment of New Technology,* ed. R. Bowers, A. Lee, C. Hershey (Beverly Hills: Sage, 1978), pp. 54–55.

11. See *Electronic News* (September 3, 1979), p. 67 and (November 19, 1979), p. 25 for estimated satellite preparation and launch costs.

12. RCA Government Communication Systems, *Electronic Message Service—System Definition and Evaluation, Final Report* (Washington, D.C.: NTIS, 1978), pp. 3–45.

13. Post Office Engineering Union, *The Modernization of Telecommunications* (London: College Hill Press, June 1979), p. 30. Also see Walter S. Baer, *Telecommunications Technology in the 1980's* (Rand Paper P–6275, December 1978), p. 42.

14. Baer, *Telecommunications Technology in the 1980's* (Rand Paper P–6275, December 1978), p. 10.

15. Frederic G. Withington, ''Future Computers and Input-Output Devices,'' *Technology Trends* (New York: IEEE Press, 1975).

16. D.R. Doll, ''Multiplexing and Concentration,'' *Proceedings of the IEEE* 60(11):1317.

17. See Kalba Bowen Associates, *Electronic Message Systems,* for a survey of private systems.

18. See RCA Government Communications Systems. *Electronic Message Service* for more details.

19. Commission on Postal Service, *Report,* vol. 2 (Washington, DC: U.S. GPO, 1977), p. 439.

20. S.J. Lipoff, ''Mass Market Potential for Home Terminals,'' *IEEE Transactions on Consumer Electronics* (May 1979), p. 47.

Appendix 2A: General Technical Aspects of Electronic Message Communications

Every electronic message service will employ a common set of engineering techniques in the design of new systems. This appendix discusses available procedures for ensuring that messages are completely, efficiently, and accurately conveyed to intended recipients without unnecessary delay.

Message Coding

All messages that enter an electronic message system must have their contents translated into some form of machine-readable code. The basic unit of information in digital processing and electronics is the binary digit or bit, which can be assigned the property of one or zero at any given time. Machine language consists of streams of bits, forming strings of ones and zeroes. Messages traveling through an electronic system are transformed into machine language by applying one of several basic translation techniques.

Character encoding can be accomplished by first specifying a unique binary identifier code for each character in an alphabetic or numeric set (alphanumerics). Identifier codes must be also specified for spaces between words, line spacing, punctuation symbols, and special symbols. Messages composed of alphanumerics and other symbols can then be serially encoded, character by character, according to a chosen code scheme. This process transforms written language into streams of binary code that a machine can read and use to reconstruct the contents of an original message.

Various coding schemes have been developed for use in domestic and international message-transfer systems. Each assigns different identifier codes and different bit lengths to each character. For instance, several versions of the Baudot code, carried primarily on Telex systems, use five bits and a shift signal to specify the alphanumeric set and special symbols or functions. The ASCII code, which is very popular in domestic applications, uses a bit length of seven.

A second available translation technique, called image coding, can be used for messages containing graphic information or for messages lacking alphanumerics. Images contained in a message may be encoded by dividing each page into a uniform matrix or grid of very small cells called pixels. The contents of a message causes particular pixels to be shaded and others to be clear. By sequentially marking a one for each pixel that has shading and a zero for each without, the contents of a page can be digitally encoded into a machine-readable language. This procedure yields a stream of binary bits equal to the number of pixels on a page and permits page contents to be recovered after decoding.

Although both of these techniques will recover the contents of a message, the decoded product will rarely match the original document. Character encoding will yield an identical decoded message only if exactly the same kind of paper and font (typeface) are used at the receiving end as originally used to construct the document. Image encoding will result in decoded output with lower quality of resolution than the original unless a very large number of cells per inch are used to encode the originating message. The currently available technology of electronic message transfer emphasizes the transfer of message contents (that is, information) over the quality and resolution of exact document reproduction. However, very costly equipment could be developed that would primarily emphasize reproduction quality.

Bit Requirements for Messages

Each possible message-encoding technique can generate computer information in varying numbers of bits as a result of several factors. When character encoding is used, the number of bits per message page depends directly on the number of characters on the page and the encoding scheme employed. Several studies of characters per message page reveal that the maximum number is around 6,600 characters with typical values ranging from 1,800 to 3,000 characters. Assuming the use of an 8-bit character code, each page would typically generate between 14.4 kilobits and 24 kilobits of machine information.

In image encoding, the number of bits per page depends on the resolution quality required of the message to be encoded. The number of generated bits can be roughly determined by using the formula:

$$N = (LW) \times (R)^2$$

when N is the number of bits, LW is the area of the page in square inches, and R is the resolution quality desired in elements per inch. Therefore, an $8\frac{1}{2} \times 11$ page requiring 100 elements per inch would generate about 935 kilobits. For the same size page requiring a higher resolution quality of 200 or 300 elements per inch, the number of bits produced would correspondingly be about 3.8 or 8.4 megabits. It is apparent from these comparative calculations that image encoding that depends only on page size and resolution quality has much higher bit requirements than character encoding. Without special techniques to reduce this difference, character encoding can be more than 100 times more streamlined than image encoding.

The preceding discussion derives the magnitude of bits generated by various message encoding techniques to be handled and conveyed by a transfer system. These figures include the bits resulting solely from the translation of message contents into machine code. The actual transfer process requires the addition of supervisory information with each message. Supplementary machine code is necessary to perform error-checking activities, communicate system status and timing information to all equipment, and ensure that messages can be properly reconstructed at the receiving end. All those activities generate overhead bits that must supplement the basic message bit requirements. Consequently, bit requirements for messages are larger than the preceding calculations indicate.

Data Compression

Various techniques compress or reduce the data bits required without reducing the clarity of the message or introducing errors in the translation process. These techniques are particularly useful when applied to image encoding because the generated machine code requires more bits per message page than character encoding. One way to compress a bit string of ones and zeroes is to record only changes in strings instead of copying each one and zero as it occurs. This technique has been shown to reduce the bit requirements of an image-encoded page with a resolution quality of 250 elements to an inch by a factor of nine.[2] Thus, a document page requiring about 4 megabits can be compressed to about 450 kilobits by using this technique.

A second compression technique which is more computationally complex relies upon the fact that message page contents are patterned and do not cause a random shading of matrix cells. For instance, all the cells positioned over a blank line of a double-spaced type document will be very predictably without color. Similarly, the cells vertically adjacent to an uppermost shaded cell that begins a letter character such as "I" are also likely to be shaded. The implication is that shading and the absence of shading is clustered in a somewhat predictable, nonrandom manner. This second technique attempts to correlate the binary strings of adjacent lines as well as horizonally adjacent elements and records changes only if predicted outcomes (that is, the presence or absence of shading of a cell) are not met.

This second compression technique, which is a two-dimensional encoding scheme, reduces message-page bit requirements by eleven times the uncompressed amount.[3] Although this method achieves the greatest compression, it also is more costly and difficult to implement and it produces more translation errors than the former technique. Future compression techniques will allow data to be reduced fifteen to twenty times the uncompressed requirements.

Channel Capabilities or Bit Rates

Messages encoded into bit streams are transferred along channels with capacity limits that are measured in bits per second. Various types of transmission channels are designed to accommodate different message-transfer capacities. Narrow-band channels are designed to carry up to 300 bits per second, whereas voice-band systems transfer between 300 and 9,600 bits per second. The choice of channel alternatives will critically influence the success or failure of an electronic message-transfer system.

The selection of channel size must be guided by such factors as average message size and mean number of messages to be processed per unit of time. For instance, a channel with a capacity of 4,800 bits per second could transmit 600 characters per second. If the average number of characters per page is 2,400, each page would require 4 seconds of channel time. Assuming that the channel is never idle, it could accommodate more than 21,000 character-encoded pages in a 24-hour day. If daily message volume exceeds this amount, or the average message size increases, then a larger channel size must be used or another channel must be added. Similar volume limits can be derived for different size channels.

The number of message pages that can be transferred on a 4,800 bit per second channel will be much less than the preceding calculation if the messages are image encoded. Assuming that a high-resolution image-encoded page requiring four megabits can be compressed to 400 kilobits, each page would require almost one and a half minutes of channel time. The identical, fully loaded channel operating over a 24-hour day could transfer slightly more than 1,000 image-encoded message pages. It is evident that message services that transfer principally image-encoded messages will require many more channels of a given size or much larger channels than services transferring an equal number of character-encoded messages. Without data compression, each image-encoded message page would require even more channel time and thereby further restrict the number of pages that could be transferred.

Electronic message systems may utilize a combination of different channel sizes when message volume varies or when low-capacity local channels are allowed to interconnect with high-volume long-haul transfer channels. Differences in channel-bit transmission rates may be handled by installing buffer storage devices between channels with uneven capacities. Buffer devices allow machine language arriving from high-speed channels to be stored until lower speed lines can accommodate it. Alternatively, buffer devices accept data bits from a number of low-capacity lines so that they can be combined for transmission over high-capacity links.

Transmission Channels

Electronic messages that have been digitally encoded may be transferred over either of two types of transmission channels. Analog channels require that transmitted signals be linearly modulated according to a time-varying continuous-wave form. The most familiar example of an analog channel is probably the local line connecting home and office telephone sets to neighborhood switching centers. Digital channels, in contrast, are designed to relay only a particular, limited set of signal values. Message signals transferred on digital channels are compared and interpreted only according to this set. Channels of this type are increasing in popularity as a result of falling costs of microelectronic circuits that allow these channels to be used economically and also because of the accelerating demand for data communications.

Because analog channels are not primarily designed to convey digital signals, the machine information must be given continuous waveform properties so that it is suitable for transmission. Modulation methods are employed that convert input data to streams of pulses, consisting of differing frequencies sent at a constant rate or uniform frequencies transmitted at varying intervals. The former method, called *frequency-shift keying*, discriminates between differing tones; the latter method, known as *phase modulation*, depends upon measurement of the phase difference or time delay between pulses. Phase modulation has been used together with amplitude modulation to convey machine information over analog channels in an efficient manner. Devices that modulate signals before transmission over analog channels or demodulate signals after reception, are called *modems* (a contraction of modulator-demodulator).

Digital channels are attractive message-transfer media for several reasons. Digital transmission of digitally encoded messages eliminates the need for modems. An equally desirable quality is that digital channels can tolerate extraneous noise. Because equipment must discriminate only among a limited set of signal values, it is simpler to discern received signals from background channel noise. In addition, digital signals can be clearly regenerated along a transmission path while reamplified analog signals reproduce both signal and accumulated noise. Although historical developments have led to the almost universal deployment of analog channels into business and residences, digital channels may be used eventually for local distribution of telephone service.

Transmission Techniques

Data can be transferred over a channel by employing various transmission techniques. Asynchronous systems operate at a nonuniform rate. A single fixed length of code, which is usually the size of a bit string of character code, is sent during each transmission interval. The transmitter begins the transfer process by sending a starting signal. A string of code follows and then transmission ends with a termination signal. In this process, the receiver is informed when a string of code begins and when it ends.

Another available data transmission method allows channels to be utilized more efficiently than asynchronous techniques. In synchronous communication, the exact departure or arrival time of each bit of information is uniformly controlled, thereby eliminating the need for start-and-stop signals. As a result, less system status information needs to be transferred, thus allowing more message information to be conveyed per unit time. The use of synchronous communication techniques in general allows more channel messages to be sent than asynchronous techniques, regardless of the channel type employed.

The use of digital channels offers the possibility of utilizing new transmission techniques. One such technique is message switching. Traditional communications systems, such as the telephone system, use *circuit switching*, a continuous linkage established between users during the entire duration of the message. In essence, the communicating parties seize and hold the communication path during their entire message transaction, even during intervals when little data or conversation is being transferred.

Message switching, in contrast, involves noncontinuous connection between users, allowing transmission channels to be more efficiently used. This technique relays encoded messages in the direction of the receiver without establishing a preset path between users. Messages originating from a sender are stored at a nearby switching center until facilities are available to move the message to another center that is closer to the destination. These storing and forwarding steps repeat until the destination is reached. Because this technique uses a channel only during actual transmissions and not during pauses, a high degree of channel efficiency is achieved.

Transmission systems that use message switching to full design capacity may encounter congestion problems. Delays in message reception can be alleviated by restricting the size of messages transmitted. This restriction requires that encoded messages be divided into *packets* of fixed bit length. Although this "packet-switching" technique increases channel utilization, it must also be remembered that each packet requires the inclusion of addressing, sequencing, and system supervisory information. As a result, packet switching is most attractive when messages can be contained in relatively few packets.

Error Detection and Correction

When data are transferred over transmission channels, errors may be expected to occur. Noise on a channel can cause bits to be lost or individual bits may assume unintended values. Various detection and correction techniques are available to reduce the errors accepted at a receiver.

A simple way to detect an error in a bit string of code is to add a parity bit. Assuming a choice of an odd parity rule, each 7-bit ASCII string of code representing a character will have an eighth bit added. This bit is assigned a value zero or one so that the total number of ones in the string will be odd. If the receiver encounters an 8-bit string with an even number of ones, it may have detected an error. This method, called a *vertical redundancy check*, is accurate only if a single bit is in error. If an even number of bits is wrong, errors will not be detected.

One method of increasing detection of multiple errors in a bit string is to add a checking bit string after a block of characters has been transferred. Each bit in this string is assigned values so that each vertical column in the block (that is, the first bits of each character code form column 1, the second bits of each character code form column 2, and so forth) observes the parity rule. Although this method, called *longitudinal redundancy check*, reduces undetected errors in the horizontal direction, it is subject to the same limitations as in the vertical direction.

An alternative way to detect errors that is less elegant but very effective involves repeating messages automatically and checking for inconsistencies. This technique, however, has the distinct disadvantage of allowing far fewer message data to be transferred than the various available methods. The most effective detector systems do not rely solely upon coding techniques and involve the introduction of hardware to address the problem.[4]

Data are generally held in buffer storage at the transmission end until the receiver has a chance to check the accuracy of transmitted bits. If no errors are detected, the receiver signals the sender to continue with a new data string. If errors are found, the receiver asks for a retransmission of the data previously received. Some error-correction equipment retransmits only one character at a time, while others resend blocks of characters or sometimes entire messages. The advantages of handling smaller blocks of data in error-correction schemes are that retransmission time will be shorter and buffer sizes can be smaller. However, it would be as inefficient to handle very small quantities of data (such as the code representing a single character) as it would be to handle very large amounts of data.

Message-Transfer Security

Methods have been developed to promote the security of message communications. The initial choice involves choosing between safeguarding the system hardware and software from malicious tampering (and therefore also safeguarding the messages in the system) or protecting the readability of the messages themselves. Although most electronic message services attempt to provide at least some minimum level of system protection, few can justify the very high cost of completely securing every data channel and terminal device from unauthorized use. Because of this economic reality, efforts have been devoted to developing cipher techniques that increase message security.

Cipher techniques alter text messages so that their contents cannot be read or modified by unintended recipients. Three basic types of ciphers have been developed. The *transposition* approach rearranges the sequences of text characters without changing their identity. The *substitution* technique replaces text characters with new symbols without changing their order. *Product ciphers* alternate transposition and substitution steps. Using such techniques, only persons with access to cipher keys should be able to unscramble received transmissions. However, a determined eavesdropper with adequate knowledge and technical resources can decipher virtually any transmission, given enough time. In such cases, the rapidly falling costs and increasing speed of computation have worked to the detriment of developing unbreakable ciphers.

Notes

1. Philco-Ford Corporation, *Conversion Subsystems for the Electronic Mail Handling Program* (Washington, D.C.: NTIS, 1973), Final

Report, p. 7–2. Also see Jackson, C.L., *Electronic Mail*, Report CSR TR–73–2 NASA Contract No. 2197 (April 1973), p. 25.

2. Philco-Ford Corporation, *Conversion Subsystems for the Electronic Mail Handling Program* (Washington, D.C.: NTIS, 1973), Final Report, p. 4–3.

3. Ibid.

4. McNamara, J.E., *Technical Aspects of Data Communication* (Maynard, Mass.: Digital Equipment Corporation, 1978), pp. 148–158.

3 Potential Traffic and Use of Electronic Message Services

The successful implementation of electronic message services depends on many factors, including technical feasibility, economic viability, consumer attitudes, and the ability of institutional and regulatory bodies to shape technological developments that meet human needs. The discussion in chapter 2 highlighted the technical features of innovative message-transfer systems now under development. This chapter addresses user needs, consumer attitudes, and economic factors in a discussion of the anticipated market for these services. The subsequent discussion considers likely users, usage potential, and the many critical factors that will influence the rate of growth of message traffic.

The Need for Message Transfer

The need to communicate over distance has been evinced by various examples of human ingenuity. Visual message systems, such as smoke signals or semaphores, have been used since the beginning of history. During the early colonial period, written and verbal messages were exchanged by an informal courier system. The founding of the national postal system provided the public with a convenient, nationally linked, ubiquitous written message-transfer service on a regular basis. The advent of the telegraph in 1864 provided users with a faster methods of distance communciation. In 1876, the development of the telephone offered the general public still another medium for communication. Innovative electronic message-transfer services now under development will offer further options.

Historically, consumers have responded favorably to new communication systems. Mail volume has continued to rise at a steady pace since the inception of the postal system. Telegraph usage grew dramatically for almost two decades until the telephone became the predominant mode for high-speed message communications. The telephone has become an essential fixture in modern society; telephones are installed in over 95 percent of U.S. households and over 650 million calls are made per day.[1]

The continued growth of domestic communications traffic has been accompanied by dramatic shifts in consumer preferences. Over the last hundred years, the increase in telephone usage has been explosive relative to telegraph and mail usage. Telegraph messages has diminished to less than 1 percent of the total number of domestic messages sent annually. The increasing preference for telephone services over mail has also been dramatic over the last four decades. In 1945, 35 percent of all messages went by post, while the remaining 65 percent were exchanged by telephone. By 1973, this disparity had increased to 20 percent by post and 80 percent by telephone.[2] In recent years, telephone use has become even more dominant.

The popularity of the telephone can be easily understood when the attributes of alternative media are compared. Table 3–1 lists some of these characteristics. The telephone system is a convenient and rapid medium that permits an immediate two-way exchange of messages. Informal messages may be exchanged without written records. Two-way interactive communication capabilities facilitate the rapid resolution of arguments, the clarification of a communicator's meaning or intention, and the initiation of informal agreements. These characteristics at least partially explain why 1.6 more adults use the telephone to maintain contact with out-of-town acquaintances than those who use the mail despite the fact that the telephone is more expensive.[3]

Although telephone usage may draw message traffic away from postal services, telephone use obviously is not a direct substitute for mail. However, the growth of telephone message volume may be responsible for reducing the growth of mail volume. It was estimated that telephone usage reduced first-class mail volume by almost 5 percent in 1975.[4] Yet the continued growth of mail volume suggests that voice-oriented communciation technologies will never fully meet all message-communication needs.

There are many reasons for the enduring nature of mail service. Written messages meet a variety of social and business uses. For instance,

Table 3–1
Characteristics of Major Message-Transfer Systems

	Mail	*Telegraph*	*Telephone*
Cost[a]	Very cheap	Moderate	Cheap
Transfer Speed	Slow	Rapid	Instantaneously
Hard-Copy Record	Yes	Yes	No
Convenience	Moderate	Difficult[b]	Easy
Two-Way Interaction	Very slow	Slow	Immediate
Transformation of Message	No	Yes	Yes

[a]In 1981 a 3-minute daytime station-to-station call between New York and Seattle cost $1.58. By comparison, a 15-word telegraph message between those points cost $7.25 while a letter cost $0.20 for roughly two pages.

[b]Requires code knowledge and access to machinery for direct use.

mail transfer promotes commerce. It allows retail sales to be announced publicly and permits residents of small towns to purchase goods and services found only in larger cities. Mail service also facilitates the transaction of business. Contracts may be negotiated between distant points; documents may be transferred among businesses, government organizations, and individuals. In addition, the mail system allows personal messages or greetings to be exchanged and bills to be paid. Finally, the mail system disseminates news, entertainment, information, and even culture nationally among households. The value of the mail service cannot be overestimated in the context of present society.

The multitude of uses of mail can be categorized according to function. Messages that travel through the mail system serve a number of general purposes, including the following:

1. *Information transfer,* either directed to specific parties or "broadcast" to a larger audience. In the former case, the message conveyed may be a personal letter or a greeting, while in the latter case the information sent may be a catalog, magazine, or advertising circular. The written record substantiates efforts to request or provide information among communicating parties.
2. *Verification,* which confirms the details of a pending agreement or contract. These documents are generally legally binding and establish a formal contractual relationship among involved parties.
3. *Notification,* which officially informs recipients of changes of status or situation or of a required obligatory action. For instance, insurance companies often use the mail to cancel policies; the government also informs citizens of their liabilities or responsibilities (such as tax obligations or jury duty).
4. *Authorization,* which allows persons to take actions on behalf of or with the permission of others. Checks, for example, are written authorizations that allow recipients to collect payments for goods or services from designated financial institutions.

This brief list of types of mail, of course, ignores the very personal message that a drawing from a child, a greeting card, or a love letter might convey. Personal messages of this type cannot be distinguished solely according to purpose or content. A love letter is a love letter and a magazine is a magazine. Delivered in another form, they may accomplish any of the above objectives but they will not be the genuine and familiar articles.

It is significant that the post is the only medium among the three which does not physically alter the contents of the communication. Telephone and telegraph messages must be transformed into electrical signals and are then reconverted to voice or paper at the receiving end. Because postal messages are exchanged in an unaltered form, mail may be preferred by those concerned with message integrity and authenticity.

This brief examination of the uses of communication alternatives gives some indication of the traditional functions of messsage systems. Many of these purposes cannot be achieved by current voice technology at a cost comparable to that of the mail. Other communication functions might be provided by voice systems but only after substantial modification of business and social etiquette or after significant alteration of legal rules regarding contracts, liability, and due process. The extent of these difficulties suggests that although voice-message traffic can be expected to continue to increase, our society will also continue to require a medium for written message transfer.

Emerging electronic message-transfer technology will undoubtedly stimulate a range of innovative services and may fulfill a wide variety of previously unsatisfied or unperceived needs. These developments will become important components of the future message market. In addition, electronic services may fulfill communication needs now provided by conventional technology. Thus, the developing electronic-message market will have a mix of new capabilities and traditional services that may draw traffic away from conventional media, especially if there are apparent advantages in terms of cost, speed, or performance.

The Movement of Conventionally Conveyed Messsages to Electronic Systems

A large portion of the anticipated message volume is likely to be diverted from conventional media. The potential growth of electronic-message traffic can be analyzed by understanding the nature of the messages conveyed and the service characteristics of antecedent technologies. Appendix 3A surveys the composition and traffic characteristics of both telephone and postal message volume. The following discussion draws upon this descriptive material.

Postal Message Traffic

Postal officials have suggested that some items that are now mailed can just as easily be conveyed by electronic message-transfer systems. Obviously, fourth-class parcels with intrinsic value (such as merchandise) are not amenable to electronic transfer. Second-class mail tends to be relatively lengthy specialized communcations; various entrepreneurial ventures are currently aimed at developing special systems for that market (for example, the videotex services mentioned in chapter 2). First- and third-class mail are clearly be the most attractive candidates for electronic message-transfer services.

Rather than describing the mailstream in terms of postal classes, mail can be classified according to content or generic type. The mailstream includes:

1. *Transaction*—commercial or financial arrangements; transaction items generally consist of checks, invoices, bills, statements of account, purchase orders, or other financial papers.
2. *Correspondence*—cards and letter messages.
3. *Advertising*—items specifying the price or availability of goods and services, as well as solicitations to buy, join, attend, or contribute.
4. *Magazines and newspapers*—publications disseminating news, entertainment, culture, or information.
5. *Merchandise*—parcels and other items not amenable to electronic transfer.

These categories are closely related to the mail classes described previously. For instance, most merchandise moves by fourth-class mail; most advertisements and periodicals travel by third and second class, respectively. However, all categories can be sent by first-class mail when mailers superior service is required. Most first-class mail consists of correspondence and transactions, a distinction made because the latter tend to be more specialized items and physically distinguishable from the former.

Transactions, correspondence, and advertisements may be the most logical candidates for future diversion to electronic systems because of message content and length. Service developers have focused on the first two categories. Postal authorities feel that advertising is an additional promising candidate, especially since the USPS hybrid system is being designed to print and process batches of messages with a largely fixed format and at least partially fixed content. Form letters, bank statements, and advertising messages also typically have these characteristics.

Although there have been year-to-year fluctuations in volume, data pertaining to the origin and destination patterns of mail reveal consistent trends, especially with respect to transactions, correspondence, and advertising. Table 3–2 illustrates these trends. A more extensive discussion of origin-destination patterns can be found in appendix 3A. These patterns will become particularly important in configuring and deploying message-transfer equipment.

Most mail, whether transactions, correspondence or advertising, flows from the more than 6 million businesses to 75 million households.

Table 3–2
1974 Mail Volumes by Sender and Receiver

	Business		*Household*		*Government and Nonprofit*		*Total*	
Receiver	*First-Class Mail*	*All*	*First-Class Mail*	*All*	*First-Class Mail*	*All*	*First-Class Mail*	*All*
Business	17.0	15.2	14.0	9.2	1.8	1.3	32.8	25.7
Household	39.6	46.0	16.1	10.9	8.5	12.9	64.2	69.8
Government and Nonprofit	0.2	1.6	1.1	0.7	1.1	0.7	3.0	3.0
Total	57.4	62.8	31.2	20.8	11.4	14.9	100.0	98.5

Source: Commission on Postal Services, *Report*, vol. 2 (Washington, D.C.: U.S. GPO, 1977), p. 926.

Interbusiness postal volume, which potentially may be carried by end-to-end electronic systems, is two to three times less than business to household mail.[5] To develop the largest and possibly the most lucrative market for electronic service, service providers must encourage households to acquire terminals or offer conventional message collection and delivery services similar to those of a hybrid system such as the one designed by USPS.

Mailing volume is, in general, not well distributed among the more than 6 million businesses. Large retailing corporations, such as Sears, as well as banks and credit card companies tend to generate large volumes of mail. In other words, a very small number of businesses generate most of the volume. A recent survey found that 86 percent of postal revenues was generated by only 10 percent of all corporations. These firms alone were responsible for transmitting almost 21 billion pieces, consisting mostly of transactions and correspondence. This concentration of message originators will be an obvious initial target for service providers.

In contrast, service providers may find that households are less likely to accept end-to-end electronic systems because they receive a very low volume of weekly mail, most of which are personal cards. Thus, more business messages will be moved from conventional mail to electronic systems, at least in the initial stages of development; however, a hybrid system will permit households without terminals to access electronic systems.

Government and business organizations may be very interested and may have the necessary capital to install local terminals for electronic message transfers. Data compiled over several years reveal that mail flowing among business and government units amounts to between 19 and 29 percent of all mail. When considering only first-class volume, this flow comprises between one-fifth to one-third of all transactions and correspondence. Equipment manufacturers and developers of new end-to-end electronic systems will concentrate on the government-to-business market along with the intracompany message market, which now travels outside of the postal system as discussed later in this section.

Mail is basically an urban phenomenon. The 75 largest post offices, all located in urban areas, account for over 50 percent of all volume and generate almost one-half of all revenue. In addition, 60 percent of all letter mail remains within the same metropolitan area. Businesses are primarily centralized in urban areas as are the consumers they serve. If developers concentrate only on the most lucrative postal markets, electronic systems could evolve as an urban phenomenon with interstate facilities developing principally among major metropoliltan areas. Rapid development of a mass consumer market would probably require concerted promotional efforts by both business and government.

Telephone Message Traffic

Although various attributes of telephone-message traffic, discussed in appendix 3A, are similar to those of the postal mailstream (such as business nature, localized traffic, and limited message size), these media may not be interchangeable for every communication need. The instantaneous, often informal nature of telephone communication in no way resembles the deliberate speed or the delayed interactive nature of mail service. Patterns of usage of each mode are intimately linked with such factors as availability of service, cost, and convenience. Message-traffic statistics fail to indicate that each communication mode meets particular message needs, yet end-to-end electronic message-transfer service providers may attempt to develop a market among telephone users for some message needs. A policy of flat-rate pricing may make electronic transfer services even more attractive for some telephone users.

Messages Conveyed Outside of the Post and Telephone Services

There are, of course, messages which are not carreid by USPS or by telephone or telegraph companies. For example, expedited messages are delivered by special couriers at premium prices. A second, much larger category—intracompany correspondence handled internally by large multiplant corporations—is considered to be a prime candidate for electronic transfer.

Messages carried by special courier are currently subject to the Private Express Statutes (PES) that require that postage be paid on such items. This stipulation suggests that some specialized message transfers may already be incorporated in reported postal statistics. In November 1979, the Postal Service exempted time-sensitive mail (that is, those items which lose value rapidly as delivery is delayed) from PES requirements in the face of mounting pressures from Congress and interest groups. As a result, time-sensitive messages may be carried by electronic service providers without challenge by USPS.

Little solid information is available on the annual volume of intracompany or interoffice mail. One report suggests that interoffice correspondence is roughly two to four times greater than postal correspondence. Another, more recent examination estimates that only one-third of all intracompany mail moves in the first-class stream. The remaining correspondence moves by courier pouches or in company vehicles. The Xerox Corporation suggests that intracompany volume totaled 44 billion pages in 1975 and may reach 72 billion pages by 1990.[6] This massive message stream will not be ignored by electronic system developers.

Beyond Conventional Message-Transfer Needs

The characteristics of messages carried on electronic transfer systems will differ from the features of conventionally conveyed messages. New technological capabilities will stimulate new uses. For instance, elected officials might use feedback channels to solicit citizen opinions concerning crucial votes about to be taken. Fortune tellers or astrologers might open 24-hour counseling services. The possibilities for new services are limitless.

Someday persons living on opposite coasts may play chess together or offer step-by-step advice regarding meal preparation or auto repair while on-line. Messages conveyed over electronic systems may include interactive entertainment or personal counseling and instruction. Human-machine communications are also possible. Although such communications do not fall within this study's scope, these hardware and software system developments could facilitate the evolution of the "electronic cottage." The magnitude of message volumes generated from such new services and uses, while difficult to predict, are likely to be very significant.

The Nature of Electronic Message-Transfer Traffic

Although a precise estimate of the potential magnitude of electronic transfer-message volume would be highly desirable, such a forecast may be less valuable than an estimate of the scale of evolving possibilities. Long-term market forecasts of projected sales are usually inadequate and often wrong. The basic estimates of future message volume, system distribution, and likely user characteristics in this section, however, give some indication of the distribution and severity of the resulting effects of widespread use of electronic systems.

Aggregation of Message Streams

Electronic message-transfer volumes may include four identifiable sources of person-to-person message traffic. Electronic systems are expected to attract traffic from the postal mailstream and intracompany transfer systems. In addition, telephone messages could be diverted to electronic systems under certain conditions. These message streams, together with volume generated from possible new uses, may be aggregated to produce an estimate of the potential volume of electronic message transfers.

The mailstream is generally regarded as a principal source of potential electronic message traffic. First-class mail (that is, transactions and correspondence), in particular, may be diverted to electronic systems because these messages are often concise, alphanumeric, and in a standard format. In addition, because most first-class mail is of a business nature, companies may switch to electronic transfer to save time and money. However, many first-class business messages are candidates for electronic funds transfer (EFT), an issue which will be treated later in this chapter.

The volume of first-class mail could potentially reach between 60 billion and 75 billion pieces before the end of the decade. Postal officials expect check-related mail items to number almost 45 billion pieces and noncheck items to be between 15 billion and 30 billion pieces. Electronic systems may displace between 17 billion and 20 billion messages from the total volume of first-class mail. Of these diverted messages, EFT might account for up to 6.5 billion items. But slow consumer acceptance of EFT has raised doubts about the rapid growth of EFT volume over the next decade.

RCA has made a separate projection of the market potential of the hybrid system described in chapter 2. Table 3–3 reveals that almost 35 billion first-class items are considered candidates for electronic transmission. This projection excludes transaction mail that may be included in EFT systems. Thus, between 11 billion and 35 billion items may be

Table 3–3
Projected Number of USPS Messages with Electronic Transfer Potential
(billions)

Source and Letter Category	*First Class*	*Third Class*
Business Mail		
Transactions	18.8	
Correspondence (Individual)	8.7	
Priority (Alphanumeric)	0.4	
(Graphic/FAX)	0.8	
Advertising-Alphanumeric		5.7
(Graphic)		7.3
Household Mail		
Business Correspondence	3.0	
Other-Household Correspondence	3.1	
TOTAL	34.8	13.0

Source: RCA Government Communications Systems, *Electronic Message Service, Executive Summary* (Washington D.C.: NTIS, 1978), p. 52.

diverted from first-class mail to electronic message systems, depending upon implied assumptions.[7] If transactions are carried on these systems, the first-class contribution could be even larger.

Intracompany messages traveling outside of the mailstream may also migrate to electronic message services. Assuming that one-third of the projected 70 billion pages continues to move in the first-class stream and also that more than one-half of the remaining volume continues to move via conventional company mail services, then about 20 billion messages may travel over electronic systems. Of course, if electronic services become much cheaper than conventional systems, the volume of electronic intracompany messages will be somewhat larger.

Electronic message services are not likely to be direct substitutes for any current message services because the products are dissimilar. However, electronic services combine some of the characteristics of both mail and telephone services (for example, hard copy and speed), suggesting that some consumers may also view electronic message systems as desirable alternatives to the telephone for certain communciation needs. Because the volume of telephone messages is expected to reach 465 billion calls per year before the end of the decade, a mere 1 percent migration to electronic systems could represent a potential volume of more than 4 billion messages. If electronic message and telephone service rates ever reach parity and modal patterns of use shift, then the diversion of even a few percentage points will result in a significant message volume.

Thus far, these estimates indicate that the annual volume of electronic transfers diverted from conventional systems could reach between 35 billion and 59 billion messages by about 1995. But if new uses prompted by new technological capabilities raise message volumes by 10 percent, total potential traffic carried on end-to-end and hybrid systems may reach 65 billion communications annually by the end of the decade. This volume is hardly trivial. In terms of scale, this estimate is equivalent to the current first-class stream.

These estimates of volume are based on somewhat conservative assumptions and do not include all potentially divertable messages. The actual volume by 1990 may be somewhat higher or lower, depending on public acceptance and other influential factors to be discussed later. However, the estimates provide a sense of the potential scale of message traffic.

The Growth of Terminal Equipment

The character of electronic message services will be critically influenced by whether services are end-to-end or hybrid. Thus it is desirable to assess the potential growth of terminal equipment and to ascertain the public visibility of end-to-end services. Of course, hybrid services will allow customers without terminals to gain access to electronic services.

The two distinct markets for message terminals are businesses and households. The business community may accept moderately priced terminals if such equipment either facilitates business operations or generates new business opportunities. These units could be well integrated into the office environment and also serve multiple functions (such as message transfer, teleprocessing, file manipulation, and storage). Households, which could be a wider market than businesses, will accept terminals if they meet two critical criteria: ease of operation and low cost (under $1,000).

Chapter 2 contained a discussion of attempts to develop a home terminal. It also revealed that concentrated efforts over the next decade might produce home terminals at a reasonable consumer market price. In the immediate future, terminal sales in the business market are expected to be greater than in the home market. Although both estimates of the number of units now in operation and of projected sales vary widely, certain trends are clearly evident.

Increased sales of four general categories of terminals (that is, facsimile terminals, word-processing terminals, communicating word processors, and data terminals) suggest that machine hardware is becoming more prevalent.[8] The number of facsimile units has been increasing at a rate of 15 percent annually. In 1975 there were approximately 120,000 units. By the end of 1980 this volume reached 220,000 units. If the same rate of growth continues, the market will contain almost 500,000 units by 1985. Present trends indicate that three-quarters of that equipment might be leased, while the remainder is purchased. Fewer than 10 percent of these units are dedicated systems; the vast majority are used with dial-up services.

Communicating word processors represent a second category of terminals capable of supporting message transfer. (Stand-alone word-processing equipment is also of interest because these units could be converted to communicating units at an additional cost of between $1,000 and $4,000.) The number of communicating word processors has been very stable over the last 5 years, remaining at a level of about 15,000 units. Some project that this market will begin to increase to a point where between 50,000 and 150,000 units can be expected by the middle of the 1980s. However, this market for equipment remains very uncertain.

The number of word-processing units without communicating options have been increasing at a fairly stable rate. Some sources assert that the annual growth rate has been around 25 percent, but compiled statistics actually suggest a rate of 15 percent per year. In 1975, there were 200,000 such units. Four years later, that number had increased to 350,000. If this trend continues, businesses could have about 800,000 units by 1985. However, this projection is uncertain because 75 percent of those units are leased and could be returned at any time. Still, these stand-alone units represent a substantial volume of equipment that can be easily converted for terminal use.

Most message terminals are data terminals originally developed for remote computing uses. These units, which include both hard-copy and cathode-ray display equipment, continue to be used principally for remote computing and data processing. They can also be used to transfer messages. In 1975, there were about 1.4 million data terminals. Estimates for 1980 range from 3 to 5 million, representing an annual growth rate of more than 20 percent. By 1985, more than 10 million data terminals could be in use.

Although message transfer is not their primary purpose, some attempt has been made to estimate the number of data terminals which could be used for message communications. A. D. Little estimates that in 1975 about 955,000 data terminals were dedicated to message communications. By 1980, this figure reached 1.45 million. If the current annual rate of growth is sustained, terminal units will number just under 2 million by 1985.

Estimates suggest that there were over 1.5 million terminals dedicated to message-transfer uses in 1981. More than 4 million units of additional hardware may be available that can be easily adopted for message transfers, bringing the total number of potential message-transfer terminals to about 6 million. If present trends are sustained until 1985, the number of business message terminals may double while the number of modifiable units may reach 10 million. Some analysts have predicted that more than half the Fortune 500 companies will have electronic message-transfer facilities in the next few years. Over the next decade, most businesses may have message terminals offering end-to-end services; home acceptance is expected to be far lower.

Potential Users

Because a large majority of the anticipated message volume is expected to be diverted from the postal mailstream, future users of electronic message systems, particularly hybrid systems, may closely resemble those currently using conventional transfer technology. Although mail flows principally from businesses to households, some households receive more mail than others. White households receive more items than nonwhite households. Also a strong correlation exists between income and education and the amount of mail received.[9] If these trends continue, educated, white, high-income households will receive more mail, electronically conveyed or otherwise, than other households. However, it would be unreasonable to conclude that electronic services users will have the same demographic characteristics as current conventional service users. Electronic message transfer may actually facilitate the transfer of transactions or correspondence for those who find present conventional mail services undesirable (for example, because of the effort required or slowness of service). A detailed market study may reveal a radically different user group for electronic message transfer.

Certain businesses have taken the lead in acquiring message terminals. The manufacturing industry uses facsimile equipment to update prices and inventions, while legal firms, trucking companies, and the oil industry use facsimile systems to file reports, contracts, and bills. Quantum Sciences Corporation expects manufacturing firms, government agencies, and insurance companies to be innovative users of terminal equipment. Table 3–4 lists some of the dominant users of facsimile systems, and table 3–5 shows the current distribution of data terminal users. Eventually, terminals are expected to proliferate throughout the business community.[10]

Crucial Growth Factors

Many conditions will influence the growth of message traffic on electronic systems. The discussion in this section will consider several important factors that could stimulate or stifle increases in the volume of electronic message transfers.

Table 3–4
Current Facsimile Users

Sector	*Percent Facsimile Users*
Manufacturing	26
Government	18
Communications	10
Financial and Banking	10
Printing and Publishing	8
Transportation	6
Utilities	6
Retail/Wholesale	6
Business Services	5
Insurance	2
Others	3

Source: Mitre Corporation, *The Impact of Telecommunication on Transportation Demand Through the Year 2000* (Washington, D.C.: NTIS, November 1978), p. 70.

Table 3–5
Current Data Terminal Users

Sector	*Percent Data Terminal Users*
Manufacturing	23
Banking and Financial	16
Government (State and Local)	12
Education	8
Insurance	7
Transportation	6
Utilities	6
Government (Federal)	6
Wholesale	5
Retail	4
Medical/Health	4
Other	3

Source: Mitre Corporation, *The Impact of Telecommunication on Transportation Demand Through the Year 2000* (Washington, D.C.: NTIS, November 1978), p. 71.

Costs of Message-Transfer Services

For end-to-end electronic services, the costs of transferring image-encoded messages will tend to be higher than those of character-encoded messages. Costs will also vary depending on terminal characteristics. Very slow, inexpensive terminals will incur larger transmission costs than high-speed, more expensive terminals. Typical monthly hardware rental costs are expected to eventually fall to half of the current charges. Future message-page costs clearly could move below one dollar. The actual cost will depend upon future transmission costs and utilization rates of chosen hardware.

Several analyses have estimated the total costs of character-encoded messages, including transfer, terminal, labor, and associated processor costs. The Yankee Group has estimates costs of between $0.67 and $0.75 per message, with terminal and transmission costs representing 45 and 30 percent of total cost, respectively. Various private intracompany systems have estimated current unit costs of between $0.20 and $0.50, although Hewlett-Packard suggests that costs reach up to $4 when labor for message preparation is included. Panko estimates that current costs are about $0.80 per message and will eventually decrease to between $0.25 and $0.50 due to cost-performance trends in the electronics industry. Cost estimates of character-encoded messages in appendix 3A match these figures closely, allowing us to conclude that page costs are now commonly under $1.[11] Future page costs may drop by a factor of ten below current costs, especially if a mass consumer market develops.

The hybrid system described in chapter 2 could generate new costs and share some conventional postal costs. Contractors estimate that a system handling 25 billion messages annually could process and transfer messages at a cost of $.018 each. In addition, there could be common and joint costs shared with conventional activities. For instance, common costs could include labor costs for carriers who walk routes to deliver both electronically and conventionally transferred mail. Joint costs could include plant facilities, heating, and lighting costs incurred when handling both electronic and conventional traffic at different times of day. Estimates indicate that these shared costs may currently be between $.07 and $.09, bringing the total per message cost to under $0.11 per message.

Obviously, users will be influenced by the prices of competing services. Costs of alternative services may be compared in a variety of ways. Some customers may compare electronic transfer costs to the costs of conventional mail. Others may view electronic transfer as a medium for rapid communication and compare it to telephone or telegraph costs. A third group might compare the total life-cycle costs of a message. In this case, the costs of message generation, editing, storage, and transfer via communicating word processors might be compared against comparable costs in conventional systems. Each comparison, while valid, may lead to different conclusions about the most cost-effective medium.

Various consistent cost trends can be observed. Costs of electronic message-transfer technology have been dropping dramatically at a rate of 7 percent per year, while rates for conventional mail have been rising. As the cost differential between these modes decreases, one might expect to see a sustained increase in the volume of electronic system messages. Furthermore, if USPS can successfully develop its hybrid system at projected costs, the resulting service may be quite popular, if other conditions do not inhibit growth. Correspondingly, if the relative cost of electronic message-transfer services does not become competitive with alternative message transfer options, the growth of message volume could be stifled.

Capital Requirements

The success of electronic message-transfer plans will also critically depend on the ability of the financial sector to meet the capital requirements of the industry. IBM and its partners will invest approximately $425 million to develop the SBS network; each of the SBS user stations will cost $325,000. AT&T and EXXON are investing many millions of dollars in their anticipated system developments. It has been estimated that by 1986 the industry will require $6.5 billion for communication facilities and an additional $13 billion for terminal and processing equipment.[12] These capital requirements will compete with such capital-intensive projects as the development of nuclear and nonnuclear sources of energy, the extraction of resources from the oceans, and the design of efficient transportation systems.

System Access

Another factor which will influence the viability of electronic transfer services involves the ease of accessibility to systems and the cost of gaining accessibility. The Postal Service is clearly developing the hybrid system in response to these concerns; the current system design includes plans to deploy over seven thousand public electronic mailboxes and provide bulk services for large volume users. As a result, both message acceptance and delivery could be accomplished without requiring each user to acquire terminals, thereby increasing public access to the technology.

The cost of equipment and software will also affect access to end-to-end systems. The decreasing cost of terminal equipment has begun to make terminal acquisition possible for the less-affluent consumer. Data terminal costs have been falling dramatically, decreasing an average of 20 to 25 percent annually. Cathode-ray terminals are now available for under $500. Multifunction terminals with microprocessors sell for under $1,000. Costs of rudimentary message display terminals with interface equipment and software is approaching $200. Communicating word processors are expected to achieve a similar scale of price reductions, from more than $7,000 to $1,000 within a decade.

Costs are also falling for facsimile equipment. Low-speed machines are readily available for under $1,200 (or for lease at $30 per month). Faster color machines are expected to be available in the late 1980s for $3,000 to $4,000. As the technology continues to advance, color machines are expected to cost between $1,000 and $1,500 by the late 1990s. These dynamic cost trends suggest that terminal prices may eventually drop to a point where equipment acquisition costs may no longer be a barrier to access to end-to-end electronic message-transfer systems. However, until that time terminal costs will continue to influence electronic message composition, volume, and the user population.

Specialized Systems

It is possible that specialized systems will be developed to handle specific types of message exchanges and thereby inhibit the growth of more general electronic message traffic. Electronic funds transfer (EFT) systems illustrate this point.

A national commission recently defined EFT as "a payment system in which the processing and communications necessary to effect economic exchange, and the distribution of services incidental or related to economic exchange are dependent wholly or in large part on the user of electronics."[13] EFT is actually a generic term for various types of specialized systems, including: automated clearinghouse networks that permit direct payroll deposits and preauthorized payments at designated intervals; interbank clearing operations that handle monetary transfers between banks; automated banking systems that dispense cash from customer-operated teller machines; and point-of-sale networks that verify credit and transfer payments to merchants' accounts. In effect, these systems reduce the need to pass paper records or messages concerning financial transactions. The required information moves electronically; typical transactions can be conveyed in less than 5,000 bits by relatively small, secure systems. In addition to reduced paper requirements, bookkeeping is more efficient.

The implications of deploying such systems are potentially significant for USPS and developers of electronic message-transfer systems. Americans conduct more than 35 billion check and credit transactions yearly. Each check can have up to four associated message transfers including precheck items (such as bills and invoices); check transfer from writers to receivers; transfers from receiver to bank; and postcheck items (such as receipts or notices). If such messages are carried out by specialized EFT systems, a large volume of potential messages will be diverted away from both conventional and more general electronic transfer systems.

Other specialized systems could also drastically influence the volume and composition of traffic over electronic message systems. Videotex systems could have a similar diversionary effect. Although these systems may compete for the message market, they may also contribute to the overall inclination to substitute electronic media for conventional transfer methods.

Suitable Communications for Electronic Transfer

It is not always possible or desirable to encode and convey all documents and communications over electronic transfer systems. For instance, some persons may demand that original copies of personal letters be transferred or require that personal signatures rather than electronic signatures be affixed to business agreements. Table 3–6 summarizes the ease by which mail items may be encoded.

Letter mail tends to contain much more alphanumeric text than graphic information. A recent survey of business and government communications

Table 3–6
Assessment of Relative Ease of Electronic Encoding

Mail Flows				*Technological Feasibility*				
From	*To*	*Type*	*(billions)*	*Already in MachineCode*	*Character Information*	*Limited Text or Graphics*	*Extensive Text or Graphics and Color Pictures*	*Intrinsic Value*
Business, Nonprofit, Government	Business, Nonprofit, Government	Transactions	6.6	Most	Some			
		Correspondence	4.5	Some	Most	Some		
		Solicitations	3.7			Some	Most	Some
		Books/Magazines	3.1			Some	Most	Some
		Packages	0.3				Some	Most
			18.2					
	Households	Transactions	22.4	Most	Some			
		Correspondence	2.4	Some	Most	Some	Some	
		Solicitations	17.7			Many	Many	Some
		Books/Magazine	9.1			Some	Most	Some
		Packages	0.9				Some	Most
			52.5					
Households	Business, Non-Profit, Government	Transactions	6.7		Most	Some		
		Correspondence	0.6		Some	Most	Some	
		Response to Ads	1.3		Most	Some		
		Books/Packages	0.2				Some	Most
			8.8					
	Households	Letters	3.8		Some	Many	Some	Many
		Greeting Cards	4.8			Some	Most	Some
		Packages/Miscellaneous	0.8				Some	Most
			9.4					
Total			89	29	14.9	11.4	29.6	4.1
Percent			100	32	17	13	33	5
				49				

Source: Commission on Postal Service, *Report*, vol. 2 (Washington, D.C.: U.S. GPO, 1977), p. 890.

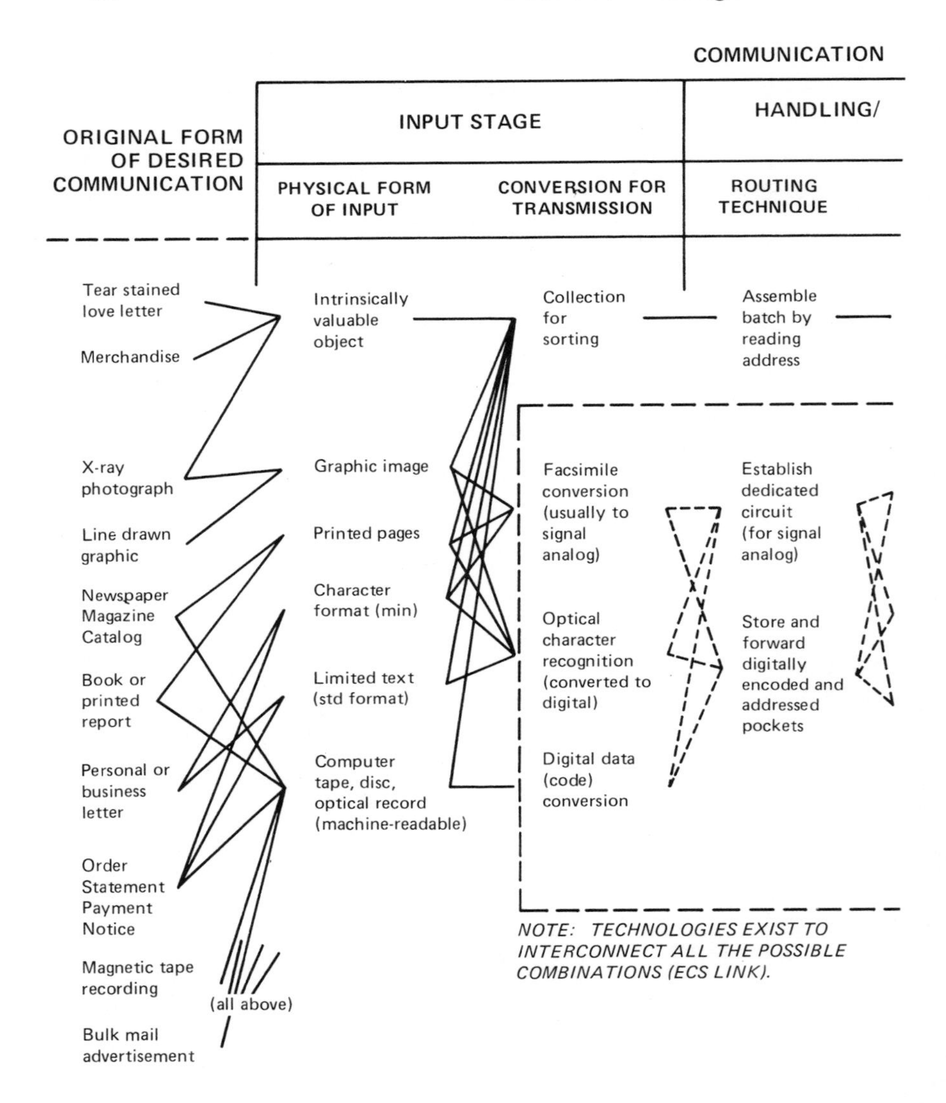

Source: Commission on Postal Service, *Report, Vol.*, 2 (Washington, D.C.: U.S. GPO, 1977), p. 882.

Figure 3–1. The Range of Alternative Communication System Configurations

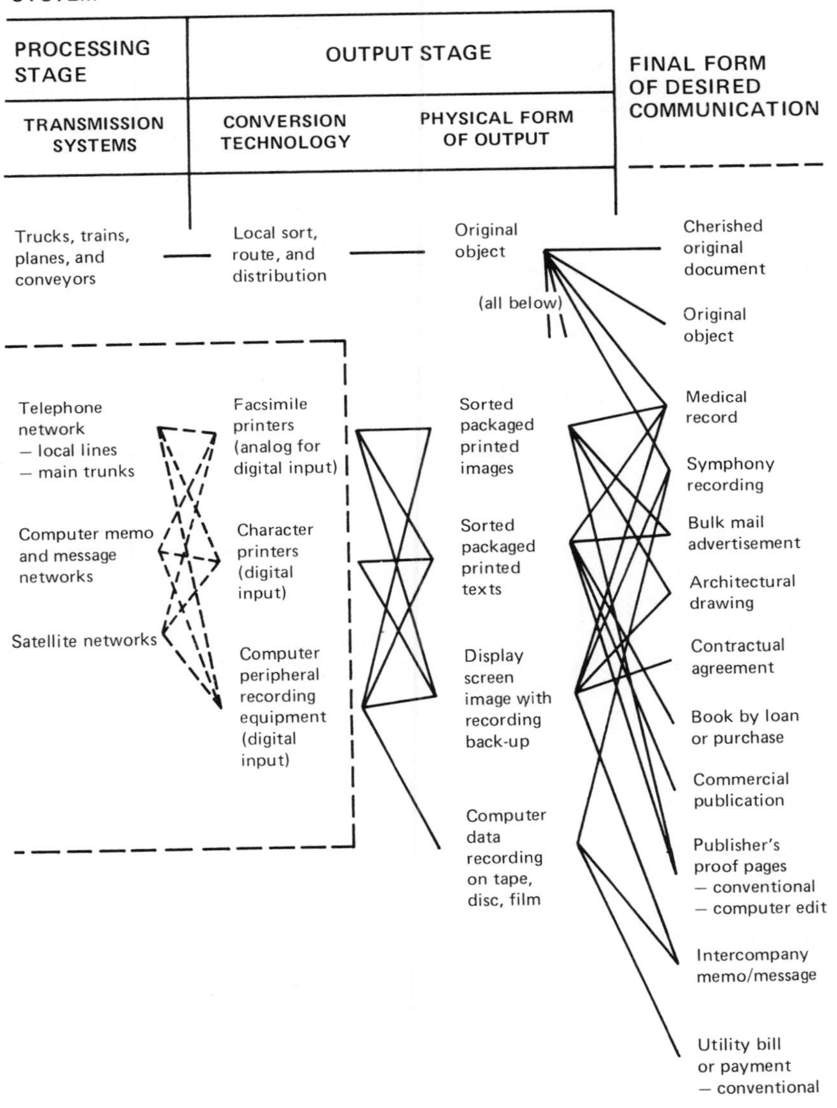
SYSTEM
PROCESSING STAGE
OUTPUT STAGE
FINAL FORM OF DESIRED COMMUNICATION
TRANSMISSION SYSTEMS
CONVERSION TECHNOLOGY
PHYSICAL FORM OF OUTPUT
Trucks, trains, planes, and conveyors
Local sort, route, and distribution
Original object
(all below)
Cherished original document
Original object
Telephone network
— local lines
— main trunks
Computer memo and message networks
Satellite networks
Facsimile printers (analog for digital input)
Character printers (digital input)
Computer peripheral recording equipment (digital input)
Sorted packaged printed images
Sorted packaged printed texts
Display screen image with recording back-up
Computer data recording on tape, disc, film
Medical record
Symphony recording
Bulk mail advertisement
Architectural drawing
Contractual agreement
Book by loan or purchase
Commercial publication
Publisher's proof pages
— conventional
— computer edit
Intercompany memo/message
Utility bill or payment
— conventional

found that letters were composed of 90 percent alphanumeric characters and roughly 10 percent graphics. The graphic content of letters typically included signatures, business logos, or specialized forms. The high alphanumeric content of letter mail makes them extremely attractive candidates for electronic transfer.[14]

Much business-originated mail is, in fact, produced by data-processing facilities. Routine business correspondence with standard formats (that is, with only a few changes to apply to a specific situation) is increasingly being produced by computer. (For example, collection agencies use standard formats for letters demanding payment of outstanding debts.) Nearly all letters involving transactions with banks and larger corporations are now created by computer. The growing dominance of machine-produced mail suggests that a transition to new transfer systems may be unnoticed by less-attentive members of the public.

Advertising, magazines, and newspapers are extensively duplicated and widely distributed. Such mail is more public in nature than letters, less personal, and exhibit a ''broadcast'' quality. They represent almost 37 percent of mail handled by the postal system. These items may also be suitable for diversion to electronic systems. Specialized information-retrieval systems offering news and entertainment services may also include advertising messages. Figure 3–1 illustrates the suitability of various types of mail for electronic transfer. If the general public reacts unfavorably to the encoding of messages, the volume of electronic service volumes could be severely limited.

Consumer Attitudes and User Expectations

The success of electronic transfer systems is critically dependent upon the degree of consumer acceptance. Such attitudes will be shaped by both experience with antecedent technology and public perceptions about the new technology. These expectations, customs, and conventions will undoubtedly influence the popularity of the new technology.

Sending messages by mail often serves as a primary and sometimes essential method of linking those who wish to communicate. Those who dispute this claim suggest that the telephone is really the primary communication medium in this country because it exists in almost all households and accounts for the vast majority of domestic message traffic. But since no one is required to have such service, the telephone cannot dependably function as a sole means of communications. On the other hand, people must routinely supply a mailing address as a condition for participation in modern society (for example, when applying for a job, establishing residency, or taking insurance policies). And messages sent by post have continued to allow parties to maintain contact with each other. Similar expectations regarding electronic message transfer would be likely only when everyone has access to a terminal.

In some situations, people are legally required to notify, communicate with, or respond to others. Such correspondence might include draft induction orders and appeals, notifications of inheritance or tax liability, and warnings about utility service termination. Telephones are unsatisfactory in these situations, because contact may not be made (for example, no telephone or no response to calls) and no written record of attempts to fulfill responsibilities exists. Mail communications sometimes are used to meet legal requirements. Postmarks affixed during the transfer process often serve as evidence of punctual or nonpunctual compliance with obligations. Again, electronic systems would have to be universally available to serve such purposes.

Once a message enters the postal system (either deposited in a street box or at a local office), users have a reasonable expectation that the transfer will be completed. Of course, users may be unaware of mistakes since nondelivery of mail that is not expected may be unnoticed by either senders or intended recipients. As a result, USPS has developed options to verify the completed transfer in writing. Certified and registered mail confirm that a piece was sent and provide a means to trace lost pieces. A sender who is worried about the delivery of a particular piece may request a notice of receipt. Because little mail is lost, these services are not widely used, however. User confidence will be an important factor in the successful implementation of new message-transfer systems.

Because mail is much slower than telephoning, many commentators have suggested that mail is not longer a medium for urgent communications. A survey of postal users has overwhelmingly concluded that a strictly guaranteed delivery date is more important than overall speed.[15] Patrons are apparently willing to mail earlier if they can be sure that a desired delivery date will be met. Urgent communications requiring a more rapid transfer speed can be dispatched by telephone or by special courier. The general public may be unwilling to pay a premium for rapid electronic handling of mundane and ordinary items that typically travel in the current daily stream.

It is not clear that consumers desire the rapid service attainable from electronic systems. Experiences with EFT systems indicate public reluctance to use such speedy systems, perhaps because few want to pay creditors so quickly. Various experiments with facsimile services have been tried by postal authorities in Sweden, the United Kingdom, and the United States. All of these trial offerings generated little message traffic, which might suggest an unfavorable consumer reaction. However, the Mailgram service has become quite popular; message traffic has been growing at a rate of 30 percent annually. Public attitudes concerning rapid message-transfer capabilities are unclear.

Because transfers by post are orders of magnitude slower than by telephone, the mail medium has a unique message-cancellation capability. It is possible to stop delivery and retrieve an item before it reaches its destination. But obviously it is fairly difficult to determine precisely when ownership is transferred between sender and receiver. Once inside the recipient's box, it surely must be the receiver's property. Postal officials are willing to attempt to stop delivery of mail until items are actually in the recipient's box. The question of ownership of messages traveling by electronic systems has yet to be settled.

One might suspect that consumers may have an aversion to using message terminals. However, experiences with an emergency locator system operated by the Stuckey Candy chain of shops reveal a counter intuitive result. Travelers exhibited no hesitancy or problems in using video terminals to contact their homes for emergency messages. Furthermore, the system was widely accepted; more than 1 million queries were processed during its operation.

Mail users feel that the post is a convenient, enjoyable, necessary, and important communication medium, although they do have complaints. Table 3–7 summarizes both the major reasons users consider mail delivery very important and the most important items transferred. These responses indicate that mail is considered important for paying bills, for conducting personal business, and for maintaining personal contacts. If electronic transfer services are to attract mail users, they must assume comparable importance in daily life and also address some of the continuing complaints about conventional mail service, such as those listed in table 3–7.

A wide variety of social and technical factors will influence consumer reactions to the new technology. New system designs must incorporate features that are easy to use and inspire user confidence in the ability of the equipment to accomplish desired functions. These features include:

1. *Reliability and dependability of service:* Messages must not be lost or misdirected and must reach destinations within specified intervals.
2. *Variety of message format:* Messages must not be so rigidly standardized in appearance so that they seem unattractive or overly uniform and depersonalized to the recipient.
3. *Equipment design:* Terminals must be designed that are simple to operate and convenient to use.
4. *Privacy and message security:* Services must provide a reasonable level of message security to reassure the user that contents will not be revealed to all but the most determined eavesdropper.

Table 3–7
Attitudes of Mail Users

Major Reasons for Rating Mail Delivery as an Important Service

Senders	*Percent Response*	*Receivers*	*Percent Response*
Necessity/Important Method of Communication	24	Important Personal Business Transactions	27
Personal Business Affairs/Paying Bills	24	Basic Form of Communication	23
Personal Communication With Family & Friends	12	Method of Personal Contact With Friends	17
Convenience	11	Enjoy Getting Mail	10

Most Important Item Sent Through Mail

Item	*Percent Response*
Payment of Bills	57
Personal Letters	26
Bank Deposits	7
Parcels	4
Mail Orders	2
Greeting Cards	1
Other Items	1
All Equally Important	4

Most Mentioned Complaints
Delivery too slow
Rates too high
Damage or misdelivery
Slow service at Post Office
Inefficiency
Delivery at wrong time of day
Discourteous personnel
Too much junk mail
Delivery to wrong address
Properly addressed mail returned as nondeliverable

Source: Commission on Postal Service, *Report* vol. 2, (Washington, D.C.: U.S. GPO, 1977), pp. 37, 48, 57.

If electronic transfer systems can successfully incorporate such features, they will have a reasonable possibility of attracting a volume of users.

Beyond the Numbers

This chapter has argued that message-transfer services play a vital role in daily life and will continue to be needed in the future. It has also suggested that electronic alternatives to conventional transfer services in the next 10 to 15 years may eventually fulfill some of our future record communication needs. At this stage it is difficult to predict accurately whether hybrid, end-to-end, or specialized systems will be dominant. But certain critical factors could influence the volume of electronic message usage and terminal equipment sales. Unfavorable developments with regard to any of these identified factors could severely constrain growth below anticipated projections.

Although electronic services may substitute for conventional services, the products are not equivalent. While messages conveyed over alternative media may accomplish identical purposes, the wide range of such characteristics as the personalization of message formats, speed, and cost suggest that each medium offers distinctly different products. Electronic message-transfer systems should not be considered as replacements for conventional systems. Rather, they are innovative systems whose availability will increase the range of consumer choice concerning communication media. Electronic message-transfer developments will also raise new issues, some of which will be treated in later chapters.

Notes

1. U.S. Department of Commerce, Bureau of the Census, *Historical Statistics of the United States* (Washington, D.C.: U.S. GPO, 1975), pp. 784, 788, 804–806.

2. U.S. Department of Commerce, Office of Telecommunications, *The Postal Crisis: The Postal Function as a Communications Service* (Washington, D.C.: U.S. GPO, 1977), p. 67.

3. Commission on Postal Service, *Report* (Washington, D.C.: U.S. GPO, 1977), Vol. II, p. 23.

4. U.S. Department of Commerce, *The Postal Crisis,* p. 3.

5. President's Commission on Postal Organization, *Towards Postal Excellence* (Washington, D.C.: U.S. GPO, 1968), annex 3, p. 2–27.

6. Xerox Corporation, Petition for Rule Making by FCC (November 16, 1978), p. 5c.

7. Estimates from other sources fall well within this interval. See U.S. General Accounting Office, *Implications of Electronic Mail for the Postal Service's Work Force* GGD–81–30 (Washington, D.C., February 6, 1981), p. 22.

8. Commission on Postal Service, *Report,* p. 516. Also see the report submitted to the FCC by Kalba Bowen Associates, *Electronic Message Systems: The Technological Market and Regulatory Prospects* (April 1978), p. 153; Henry Geller and Stuart Brotman, "Electronic Alternatives to Postal Service," in *Communications for Tomorrow,* ed. Glen O. Robinson (New York: Praeger, 1978), p. 315; and Post Office Engineering Union, *The Modernization of Telecommunications* (London: College Hill Press, June 1979), pp. 101–102.

9. M. Kallick, W. Rodgers, and others, *Household Mailstream Study Final Report,* prepared for Mail Classification Research Division, USPS (1978), pp. 71, 95, 258, 260.

10. Mitre Corporation, *The Impact of Telecommunication on Trans-*

portation Demand Through the Year 2000 (Washington, D.C.: NTIS, November 1978), p. 64. Also see Kalba Bowen Associates, *Electronic Message Systems,* p. 165.

11. Raymond Panko, ''The Outlook for Computer Mail,'' *Telecommunications Policy* (June 1977). Also see Kalba Bowen Associates, *Electronic Message Systems, pp. 35–42, 138–139.*

12. Kalba Bowen Associates, *Electronic Message Systems,* p. 211.

13. Commission on Postal Service, *Report,* p. 441.

14. R.J. Potter, ''Electronic Mail,'' *Science* 195 (March 18, 1977).

15. U.S. Postal Service, *The Necessity of Change* (Washington, D.C.: U.S. GPO, 1976), p. 167; Commission on Postal Service, *Report,* p. 49.

Appendix 3A Market Characteristics

Mail Characteristics

USPS currently offers twelve classes and various subclasses of mail service, established according to explicit rate and content specifications. Mail can be more simply defined in terms of four major classes. First-class mail can be generally thought of as anything that fits into an envelope and that is delivered according to certain time-distance performance standards. USPS has a subclass for pieces weighing over 16 ounces that are delivered according to the same performance standards. Second-class mail generally consists of time-sensitive printed items, such as magazines, newspapers, or journals that require speedy delivery. Such items have been carried at subsidized rates to facilitate the free exchange of news, ideas, and opinions. Third-class mail includes bulk circulars, miscellaneous advertising, and small pieces. Fourth class is commonly referred to as parcel post and includes a special rate for bulk pieces, such as books, records, or library materials. These four categories account for about 95 percent of total mail volume.

Fairly consistent trends have been observed over the last three decades. First-class items represent over half the volume carried and generate the most revenue. Third-class mail constitutes the second largest volume. Second- and fourth-class items make up the smallest shares of total mail volume. These figures do not include pieces carried by private-sector firms, which have become and extremely competitive force in all but the first-class market.

Traditionally, letter mail has been most commonly associated with the exchange of messages between friends and relatives. However, statistical data reveals that correspondence, including greeting cards, business letters, and personal letters, compromises only 30 percent of first-class volume. The remaining 70 percent is composed of transactions. In fact, personal correspondence represents only 3 percent of letter volume, while cards and business letters dominate the correspondence category with 16 percent and 11 percent of first-class respectively. In recent years even fewer letters and cards have been sent through the mail each year.[1]

In terms of total mail volume, 1968 data showed that transactions represent the largest share of the mailstream comprising almost 40 percent. Correspondence is only the third largest category, amounting to slightly more than one-half of the transaction percentage. Advertising makes up 26 percent while the remaining 12 percent consists of merchandise, newspapers, and magazines.[2] These proportions have fluctuated since 1968 because of several factors. Rate increases for advertising items have stimulated the emergence of alternatives to mail advertising. Various direct deposit plans, remote banking schemes, and other innovations have restricted the growth of transactions mail. Yet, these three categories together consistently have accounted for well over one-half of the total mailstream.[3]

Mail can be thought of in terms of such physical characteristics as weight and content length. Transactions and correspondence are typically sent by letter mail, while advertising usually travels by bulk mail. Letters tend to be fairly light, averaging slightly over half an ounce each and consist of an average of almost two pages per envelope. Bulk mail pieces, which are principally third-class items, tend to be about four times heavier and average over seven pages each. These attributes suggest that first-class mail tends to contain shorter, more concise messages than third-class mail.

Mail travels among over 83 million possible points. These origins and destinations can be grouped according to three categories: businesses, households, and government and nonprofit organizations. Data concerning origin points describe the category of sender rather than the physical point where mail is collected (such as collection boxes, post offices, or company mailrooms). Households receive over 90 percent of all mail. Origin and destination points have been increasing at a rate of 2.7 percent annually because new housing starts and business ventures. The total number is expected to reach about 94 million by 1985.[4]

Mail flow between origins and destinations is not a random phenomenon. Consistent trends have been observed over time. Given the dominance of transactions in the total mailstream, it is not surprising that businesses are by far the largest source of mail. Business-originated mail has continued to make up over half of all mail volume and has sometimes accounted for as much as 75 percent of total volume (it comprises 57 to 70 percent of first-class volume). Most business mail is sent to households. Interbusiness flows have been observed to be two to three times less than business to household mail.[5]

More mail is sent to households than to any other destination. Over the last decade, households have received between 60 and 70 percent of all mail. Individual households receive an average of between ten to fifteen items per week, of which three to six items tend to be correspondence or transactions. However, the household reception rate is not uniform; Monday and Saturday are the lowest volume days. The flow of mail increases to a constant level from Tuesday through Friday. On an average day, one-quarter of all U.S. households receive no mail; 17 percent receive only one piece and an equal number receive only two pieces.

The principal flow of mail is from business to household. Businesses have over the years received anywhere from 25 to about 35 percent of all deliveries. Although households originated between 18 and 22 percent of the mailstream, almost one-half of all household-originated pieces were greeting cards or post cards.

The third origin-destination category includes state, federal, and local government units and nonprofit organizations. These organizations send three to five times as much mail as they receive. In addition, a large majority of this mail moves in the first-class mailstream at a subsidized rate.

Origin and destination patterns of mail can be further disaggregated by message type—transactions, correspondence, and advertising. Over 80 percent of all transactions are originated by business while households send 14 percent. Interbusiness transactions amount to 43 percent, while business-to-household mail total 37 percent of all transactions. Sixty percent of correspondence mail flows between households. Ten percent of correspondence moves from governments to households, while only 9 percent is interbusiness correspondence.[6]

Advertising messages understandably originate solely from businesses. Business-to-business advertising represents over 20 percent of all advertising mail. The remainder flows to households, including coupons, samples and information about something to buy, events to attend, causes to contribute to, and membership solicitations. In the future, direct-mail advertising flows will depend, in part, on the relative costs of message-transfer alternatives.

Mail does not typically move over great distances. Forty percent is distributed locally. Another 35 percent is intrastate mail. Only one-quarter of all mail leaves the originating state. Thus, mail has predominantly a local or regional charcter.[7]

The Character of Telephone Message Transfer

The telephone has become a ubiquitous instrument in modern society and is used to transfer over 80 percent of annual domestic message traffic.[8] It can be found in nearly every business establishment and in all but a number of households. Annual telephone traffic amounts to over 235 billion calls, of which about 92 percent are local (that is, within a caller's service area) while the remaining 8 percent are long-distance calls. More than half these calls originate from businesses rather than residences.

Telephone messages can be characterized by message length, calling distance, and origin and termination points. Each household makes an average of 120 calls per month. The median monthly rate is actually closer to 90 calls; the average rate is increased by a minority of households generating a larger volume. One reason for these high usage statistics might be that 90 percent of all households pay a flat monthly rate for an unlimited number of local calls.

Despite the high usage rate, residential calls tend to be directed to a limited set of recipients. One-fifth of all originating calls go to the same receiving number. Roughly 30 to 40 percent more are placed to the four next most frequently called numbers. More than half of all residential messages go to the five telephone numbers most commonly dialed by each caller.

Messages tend to be of fairly limited duration. While the average length of local calls is shorter than the mean toll call rate, measuring between 4 to 5 minutes depending upon market surveyed, the statistic is actually misleading. One-half of all residential calls last less than 1 minute. (Again the average is raised by the smaller volume of calls lasting over 20 minutes.) The very limited duration of most residential calls suggests that people tend to prefer brief messages.

Finally, the destination of residential telephone messages tends to be very localized. Seventy percent of all residential calls are within a 5-mile radius of the household. Almost one-half of all residential calls terminate within 2 miles of the originating point. These figures suggest that most telephone messages pass within the same neighborhood.

Costs of Telephone and Mail Services

Until 1863, mail rates depended on distance and the number of sheets enclosed in each letter. Later, postal rates became dependent only on weight. Although the letter rate has risen dramatically over the last decade (to 20 cents per ounce in 1982), it has been fairly stable when measured in constant dollars. The other mail classes have also experienced rate increases as a result of reduced cross-subsidization among classes.

Future conventional mail rates are difficult to predict. If volume drops as a result of electronic diversion or any other reason, prices can be expected to rise unless subsidies are provided or services are decreased (for example, by eliminating Saturday delivery). Because delivery points are growing at a rate of more than 2 percent annually and these costs must be covered by revenues that come principally from message traffic, rates will have to rise if first-class volume declines substantially. It is possible that without government intervention, a vicious cycle could develop in which an initial diversion leads to an conventional rate increase that diverts more messages and so on. Thus it is extremely difficult to predict exactly what the conventional message transfer rate will be during the next decade.

The rates set for telephone service have actually declined over time. Unlimited local residential service is still available for under $20 per month, although timed charges are becoming more common. Long-distance rates have fallen to a level where a dialed, daytime, station-to-station 3-minute call is still under $2 between any points across the nation. Measured in constant dollars, these rate decreases are even more dramatic.

The advances in electronic and communication technology described in chapter 2 suggest that telephone service costs will not increase radically over the next 5 to 10 years. Innovative loading techniques have permitted more traffic to be accommodated by existing equipment resulting in higher channel efficiencies. Service costs, however, will increase in the unlikely event that the volume of telephone messages decreases substantially from current levels.

Neither mail nor telephone service rates necessarily reflect the costs of providing such services. Price discrimination policies have allowed business telephone service to subsidize the development of residential service. Traditionally, first-class mail has subsidized the other classes.

End-to-End Electronic Message Costs

Cost estimates of electronic message systems are difficult to make because of the often specialized nature of particular configurations. When making comparisons one must consider system functions, inherent flexibility (such as the ability to handle various terminal equipment), and sophistication. In general, the cost elements consist of terminal acquisition charges (including amortized purchase or lease costs and maintenance charges), local access charges (for example, for dial-up telephone), transmission costs across networks, and processor connection and use charges. A complete cost estimate might include software and operator charges, user learning, and, perhaps, message preparation and handling. But these secondary cost constituents are often very difficult to uncover and are not commonly included in estimates.

Cost studies typically are based on differing assumptions regarding system features and characteristics. for instance, no uniform message length is used to derive transfer costs, making comparisons even more difficult. Thus, cost estimates vary over a wide range. Transmission costs for short messages range from less than one cent to over one dollar. When terminal, processor, software, and message-generation charges are included, these estimates vary from $4 to $6. These inconsistencies can be partially attributed to differences in assumptions concerning labor requirements and future equipment costs. Transmission costs, terminal equipment costs, and other associated costs will be discussed separately in an effort to derive an estimated cost per message.

Several examples will indicate the range of currently available facsimile transfer costs. SPC offers a transfer service for $0.25 per minute anywhere in the country at any time of day. Depending on the equipment used, message-page transfer costs are between $0.25 and $1.50. Transfer costs over Bell System facilities vary between $0.20 and $2.40 per page depending upon equipment, time of day, and distance covered. A Canadian service offers page-transfer costs ranging from $0.70 to $1.35, depending on delivery speed (from 15 minutes to overnight).

Other transfer services charge a large monthly flat fee for limited privileges or a small fee with per-message charges. Users of Western Union's Infocom system are allowed 26 hours of system time for $1,200 per month. Datapost charges $210 monthly fee for facsimile transfer to its Chicago center and uses the USPS delivery system. Graphnet charges a monthly subscription fee of $1 and $2.23 per message. In the future, facsimile transfer page costs may drop to under $0.10 as the costs of both transmission services and high-speed terminals decline.[9]

New services offer even lower costs. Available packet services are offer rates of $0.60 per megabit. This rate implies that high volume users are capable of realizing character-encoded page-transfer costs of between $.05 and $.25. Satellite systems have the potential to reduce transfer costs even further once the necessary equipment is in place. Satellite channels can deliver messages at a cost of under one cent per megabit, opening up the possibility of per-page costs of much less than one cent at any transfer distance. But these costs can be achieved only if channels are heavily used. Also, they do not include hardware costs, which are by no means trivial.

Terminal costs vary according to rental or amortization and usage rates. Three examples of character-encoding equipment illustrate this range. One estimate suggests that a simple terminal costing $1,500 and carrying 20 messages per day or 5,000 per year would yield a terminal charge of $0.30 per message if costs are amortized over that period. Cerf, in discussing ARPANET system costs, calculates that a terminal renting for $150 per month would add between $.08 and $0.75 per message, depending on message volume. Another calculation has shown that communicating word processors, which tend to be more sophisticated, and specialized terminals currently add between $0.20 and $1.76 to message costs, depending on utilization rate and equipment configuration. Similar estimates can be made by knowing the future cost and utilization rate of terminal equipment.[10]

It is difficult to predict whether service prices will correspond to these cost estimates. Developers may decide to use differential pricing or to charge close to the conventional letter rate at the time of offering to subsidize the development of less lucrative markets. Alternatively, market conditions may force prices to be set closer to costs.

Trade Implications of Electronic Message Transfer

Up to this point, the discussion has focused on domestic markets. Developments in electronic message transfer have the potential to produce a very large amount of industrial activity and profit for equipment manufacturers and service providers. The promise of lucrative markets is likely to attract considerable interest among foreign as well as domestic business concerns. Japanese business, in particular, has demonstrated an ability to penetrate American markets, especially in the areas of automobile production and consumer electronics. The once U.S.-dominated manufacture of television receivers now has severe competition from Japanese concerns. In a development more directly related to electronic message transfer, several Japanese companies have begun to market very fast facsimile equipment. Japanese business interests may move rapidly to develop electronic transfer equipment, especially low-cost home terminals, for the U.S. Market if there is sufficient consumer demand. Such a development could have a significant effect on foreign trade and the balance of payments between the two countries.

Electronic message-transfer developments may slow the accelerating pace of the U.S. global trade deficit, which grew from $6 billion to $29 billion between 1976 and 1978.[11] This potential contribution could be realized on two different levels. At the system level, electronic message transfer may help the United States to maintain the global surplus that it has traditionally preserved in the area of telecommunications and decrease the aggregate deficit. Between 1976 and 1977, telecommunications trade in terms of net export dollars actually rose from $1.25 million to $1.27 million. But foreign developers of electronic message-transfer systems and office-automation equipment are making concerted efforts to create home markets and gain a share of the world market, efforts which could shrink the U.S. share of the world telecommunications market.

Several developments in various countries are particularly significant. Both the West German and British governments are supporting domestic efforts to develop successful videotex systems that may include message-transfer options. The British Post Office, which pioneered videotex development many years ago, has already made agreements with twenty-five U.S. corporations to install in-house experimental viewdata systems.[12] Another British corporation, the General Electric Company (no relation to the U.S. company), has purchased A.B. Dick, a U.S. office equipment company, as part of its strategy to to move into the market for office systems. In Japan, government and industry have been conducting various two-way cable experiments, technology which could easily be used for electronic message transfer. Each of these activities and others might eventually result in viable systems that could compete for a share of the world telecommunications market.

On another level, electronic message transfer may help the United States to maintain dominance in the world microchip market. Although the United States currently holds a large share of that market (67 percent), some believe that Europe (10 percent), and Japan (15 percent) will eventually gain larger market shares. In Europe, many governments are providing aid to businesses in an effort to develop an electronic export market.[13] In Japan, the government works so closely with industry that many view such collaborative efforts as one large corporation—"Japan, Inc." One Philips executive has commented that "so long as the chip is innovative, the United States will lead. But as it becomes a mass-produced product, Japan is bound to become a major competitor."[14]

In summary, innovative developments such as electronic message transfer could enhance trade opportunities in electronics and communications. By encouraging such developments and by promoting industrial compliance with international equipment standards and protocols, government may contribute to efforts to reverse the trend of growing U.S. trade deficits. Of course, technological advantage alone will not determine trade balance. Trade agreements dealing with tariffs and nontariff trade barriers must be negotiated among the interested nations.

Notes

1. U.S. Postal Service, *The Necessity for Change* (Washington, D.C.: U.S. GPO, 1976), pp. 17, 30.

2. President's Commission on Postal Organization, *Towards Postal Excellence* (Washington, D.C.: U.S. GPO, 1968), p. 48.

3. Marc Porat, *The Information Economy: Definition and Measurement* (Department of Commerce, Office of Telecommunications, Special Publication 77–12(1), May 1977), p. 104.

4. U.S. Senate. *Evaluation of the Report of the Commission on Postal Service* (Washington, D.C.: U.S. GPO, 1977), p. 27.

5. President's Commission, *Towards Postal Excellence,* pp. 2–27.

6. U.S. Postal Service, *The Necessity for Change,* Washington, D.C.: U.S. GPO, 1976, p. 14.

7. General Dynamics/Electronics. *Study of Electronic Handling of Mail, Collection and Distribution* (Washington, D.C.: NTIS, 1970), pp. 2–40.

8. See Martin Mayer, ''The Telephone and the Uses of Time,'' in *The Social Impact of the Telephone,* ed. Ithiel de Sola Pool (Cambridge: MIT Press, 1977), pp. 225–245, for a detailed discussion.

9. F.W. Miller, ''Electronic Mail Comes of Age,'' *Infosystems* 24 (November 1977), pp. 58–64.

10. Kalba Bowen Associates, *Electronic Message Systems: The Technological Market and Regulatory Prospects,* submitted to the FCC (April 1978), pp. 139–141.

11. U.S. General Accounting Office, *United States–Japan Trade: Issues and Problems,* ID–79–53 (Washington, D.C., September 21, 1979), p. 4.

12. International Resource Development, ''Videoprint,'' *Newsletter* 1 (1980), p. 9.

13. *The New York Times* (January 29, 1980), p. D1.

14. *The New York Times* (January 19, 1980), p. D10.

4 Message-Transfer-Service-Providers

Emerging electronic message-transfer technology has the potential to offer rapid and economical record communication services to the public. Various organizations in both the public and private sector have recognized these possible benefits. Many large corporations are already using readily available equipment for intraoffice message transfer. Several firms offer facsimile or other document transfer services. In addition, the USPS has spent almost $19 million over more than two decades developing and testing various possible configurations of systems. One issue of particular concern to policymakers, which is prompted by this growing interest, is the market structure under which electronic services will be offered. This concern over industry structure is important because market structure is believed to affect the distribution, quality, and price of services.

This chapter focuses on the government's role in stimulating the development of electronic services. Electronic services can be viewed as a technological improvement and extension of conventional message-transfer services. Following this line of reasoning, government could direct the Postal Service to automate its operation and extend the postal monopoly to cover electronic message transfer. At the other extreme, electronic services can be regarded as new telecommunication services, with government encouraging an atmosphere of open competition with minimal regulation. In this case, USPS might be directed to adhere to its philosophy of noncompetition with the private sector and totally exclude itself from offering electronic services. Another alternative would be to involve USPS in one or more joint ventures with the private sector.

There are legal and historical precedents for all these positions. Throughout the nation's history, government has repeatedly reaffirmed its exclusive right to transfer letter messages physically. Similarly, Congress has relied on the private sector to provide for transfer of messages by electrical means. More recently, the USPS has cooperated with Western Union in offering a thriving Mailgram service. This chapter will examine the economic, legal, historical, and political dimensions of various alternatives for providing electronic service.

All Electronic and Hybrid Message-Transfer Services

Under the traditional regulatory mold, end-to-end electronic services are telecommunications services clearly falling within the jurisdiction of the FCC. Recent regulatory policy in this area has encouraged competitive entry into monopolistic or oligopolistic markets. Both the Carterfone and MCI decisions in 1968 stimulated competition in the terminal equipment market and the specialized communication market, respectively. Since these decisions were rendered, a variety of innovative terminal equipment has been developed and some lower-cost services have been introduced. Consequently, one might anticipate the development of a terminal equipment industry and a specialized electronic service industry for this market.

Indeed, various corporations are anxious to enter the end-to-end specialized business communications market. The more visible organizations include Exxon, AT&T, and SBS (a joint venture involving IBM), representing several of the largest U.S. corporations. The active interest of these powerful corporations could lead one to conclude that government initiatives are not necessary to ensure that services are developed in this market.

The USPS hybrid system, as planned, is not restricted to conventional input and output. It also could accept electronic input and permit electronic conveyance of output to remote terminals. However, USPS does not intend to provide electronic transfer services between its own facilities and a patron's facilities.[1] Thus, several different markets for hybrid-related services could develop. There may be a market for long-haul transfer services. Additionally, local services that interconnect users with processing offices may emerge along with a terminal equipment industry.

These potential developments indicate that electronic services can be provided as an end-to-end communication industry for patrons able to pay the price or they can be provided by an automatied postal-type industry. In the former case, competitive forces have already entered the field. In the later case, two questions might be raised. Should electronic-conventional hybrid services be exclusively provided by USPS or some subsidiary? If a competitive industry is created to supply this second and larger market, should USPS be allowed to participate?

In a sense, examination of the issue of a public hybrid service necessitates an analysis of the conventional message-transfer system. Because the proposed system uses critical portions of the conventional system, one should understand why these services have been publicly provided in the past. An assessment must be made whether the reasons for continued goverment provision are sound and whether such reasoning should be logically applied to the public hybrid proposal.

The Evolution of Policy Regarding Public Postal Services

History reveals several reasons why government has continued its monopoly in transferring letter mail. Government initially became involved in the provision of postal services because it considered such services essential for the development of the country. At that time, no other organization had the resources or capabilities to provide them. As private competitors became interested in providing mail services, Congress acted to prohibit their participation so that it could ensure that services would continue to expand and improve. As a result, the government monopoly on mail services continued.

Recent history has reflected a shift in thinking on the part of Congress. In 1970, amid the cries of concern over the inefficient operation of the postal system, Congress passed the Reorganization Act. This action changed the status of the cabinet-level postal department to that of a public corporation. In addition, Congress directed the USPS to move toward a marginal-cost pricing policy and to limit cross-subsidization of services.[2]

This shift in philosophy is in part the result of the growth of the postal system and the changing views concerning postal services. The postal system, utilizing conventional technology, has emerged as a mature industry in comparison to conditions in 1845 when it was struggling to continue providing facilities during the rapid migration to the West. The former rationale of revenue protection for a massive geographical expansion of service is no longer applicable because a stable nationwide system is now in place.

General attitudes have also changed regarding the provision of postal services. Before the Reorganization Act, Congress considered postal services to be essential public services provided by government at necessary cost, just as defense and public education were provided for the well-being of the nation.[3] This viewpoint has prevailed since the country's founding and continues to be supported by nearly all governments around the world. By 1970, however, various groups had developed the view that government should not bear the costs of providing general services. They felt that postal patrons should be charged prices that reflect the true costs of service, thus shifting the cost to users.

Another idea that is currently gaining acceptance in government is that greater economic efficiencies are possible when industries produce under competitive conditions. Very significant efforts have been made to deregulate and to encourage competition in the communications, airline, and natural gas industries. This trend has prompted both the executive branch and Congress to reexamine the monopoly privilege granted to the USPS several times over the last decade. As a result, economic efficiency has become an increasingly important criterion in governmental decision making.

Economic Issues

Economists agree that there are two situations in which competitive private markets do not necessarily lead to greater efficiencies. Under both conditions government action is often required. Those who support the government monopoly of letter services generally rely on one or both of the following economic arguments.

Social or Public Goods

In general, competitive markets efficiently supply only private goods and services. The competitive market solution fails when one of two necessary conditions is not met: exclusion must be feasible and consumption must be rival. Each condition is necessary for different reasons, and each is discussed below.

Purchase at a particular price signals demand for a good, encourages entry or exit of producers, and gives some minimal indication of the direct benefits of consumption. If people unwilling to pay the price cannot be excluded from receiving the benefits of consumption, no one would make purchases and the auction system of the market would break down. Demand would not be publicly expressed and market forces would not stimulate production. Exclusion is therefore necessary for the functioning of a viable and efficient market on the supply side.

Private goods are subject to rival consumption and are exhaustively consumed (that is, consumption by A precludes consumption by B). If goods are not exhaustively consumed, the cost of allowing one more person to receive the benefit or marginal cost would be zero. It would be inefficient not to allow benefits to be enjoyed. Yet costs would not be shared by those who benefit.

Goods that do not have both of these characteristics are social or public goods. In general, government takes steps to ensure that such goods are provided because sale is impossible (because exclusion cannot be applied) or undesirable (when consumption is nonrival). Examples of such goods and services include outer space exploration and national defense.

Decreasing-Cost Industries

A second situation requiring government action is the case of natural monopoly. In some industries where the costs of fixed inputs are high in comparison to variable inputs, production functions exhibit increasing returns to scale. In this situation, average unit costs decline with increasing output over the expected range of production; marginal cost will be less than average cost. When marginal-cost pricing is employed, losses are sustained by each producer. As a result, the number of firms in such an industry tend to diminish. Without government action, a few firms will eventually gain control of the market and set prices not subject to competitive pressures. An insufficient level of output will then be produced at too high a price if the market demand is great. These are the conditions which define natural monopoly.

Government action is often necessary to prevent an inefficient market result. In some instances, public ownership is assumed. In other situations, private ownership is subject to public regulation. Traditional utility services, such as power generation and distribution and water distribution, are either publicly or privately provided to a designated region. Losses incurred by public utilities are usually recovered from general revenues rather than from user charges. When private enterprise is regulated by a public agency, a private producer may be encouraged to supply more goods at a lower price than charged in a monopoly situation. In decreasing-cost industries, such as traditional public utilities, pure competition is for the most part not feasible and some sort of government remedy is necessary.

It is difficult to define the proper government role in regard to either social goods or the decreasing-cost industry situation. In each situation the choice between public regulation and public production will depend on the amount of control necessary and the complexity of the task. When the degree of regulation required is extensive, public production may be the most cost-efficient solution; otherwise, regulation of private enterprise may suffice.

Provision of Conventional Message Transfer Services

Many traditionalists believe that written message-transfer services, especially postal operations, are public services, because they provide a universal communications system that binds the nation together and offer services to everyone.[4] Also, message-transfer services may increase the quality of information flowing among citizens. In terms of the economic criteria described previously, message-transfer services are not public goods. Exclusion can be applied to those unwilling to pay the price, and message services may be exhaustively consumed by those willing to pay. But increasing the aggregate volume of message transfer or improving the operation of the economic market (by increasing the exchange of information regarding price supply and demand), may be considered as supplemental public goods. On this basis, a valid economic argument could be made for government intervention in the absence of adequate interest by private industry.

The message-transfer industry utilizing conventional postal technology has been characterized as a decreasing-cost industry.[5] But this characterization cannot be fully supported by the evidence derived from empirical studies. In 1962, Baratz concluded that the long-run costs for all classes of mail were decreasing with increasing output.[6] In contrast, Stevenson found that ''Quantitative studies of the scale characteristics of the Postal Service, however, have not provided any support for the hypothesis that the USPS is capable of realizing such economies.''[7] Merewitz found scale economics while Haldi did not. Perhaps this confusing situation is best illustrated by the fact that two different studies of this question authorized by the Kappel Commission in 1968 reached opposing conclusions.[8]

There are several reasons why definitive conclusions about the industry cost structure have not emerged. Message transfer is provided by a multiproduct firm that uses its factor inputs for various services to a large number of regional markets. Shared costs, both common and joint, are not easily separated. Therefore, it is difficult to attribute the true costs of each of the services provided.

In fact, the costs for postal letter services may be decreasing in particular regional markets. Merewitz, in a statistical analysis of the postal system, found that economies of scale depended on the size of processing centers.[9] Furthermore, scale economies may exist for portions of the postal production activity (such as collection, transfer, or processing), even if such economies do not exist for the entire process. For instance, it may be inefficient to allow more than one distribution company to serve a small town with only a few residents. Without more disaggregated cost data, it is difficult to render a judgment concerning conventional services based solely on efficiency.

A second problem is that actual costs of individual services are never compiled. USPS uses a highly restrictive defintion of short-run marginal cost in its analyses. Over the last 20 years, total costs have moved in proportion to the mail volume.[10] This fact should be be surprising because labor costs have accounted for more than three-fourths of the total expenditures for quite some time. Yet USPS analysts have stated that they have a very high fixed-to-variable cost ratio, a key feature in the assertion that economies of scale exist for the conventional system and a key means of justifying the government monopoly.

There is a lack of overwhelmingly supportive evidence for the pro-monopoly position based on efficiency. This finding does not, however, lead to a hasty conclusion that the entire conventional service industry should be immediately opened to competitive entry. An elaborated discussion of the many impacts of such a regulatory policy for conventional services is beyond the scope of this study. But this lack of strong evidence of natural monopoly makes it difficult to conclude, a priori, that the government should have a nationwide monopoly on any services at this time based on efficiency criteria.

Provision of Electronic Message-Transfer Services

End-to-end electronic services will probably develop under the guidance of FCC regulatory policy. The development of hybrid electronic services is not so simple. USPS has spent much effort in developing a design. Its proposed system is neither a traditional postal service nor a strict telecommunication service. It might be considered to be a technology with mixed characteristics. Because jurisdiction is not clearly defined, USPS may attempt to extend its letter mail monopoly to include electronic messages. Or the FCC may attempt to regulate the electronic transfer activities of USPS.

Cost studies of telecommunication systems, such as telephone systems, indicate that scale economies are not found in all aspects of service. For instance, Hall suggests that while intercity communication channels and multiplexed lines exhibit economies of scale, sophisticated user terminals and local distribution loops do not.[11] In regional markets with limited demand, it may be economically efficient to allow monopoly provision of communication services or of particular portions (such as, long-haul transfer communications within a state). In the largest urban markets (for example, Los Angeles, New York, and Chicago), the volume of electronic messages may be great enough to justify competitive intercity transmission services.

The lack of solid evidence concerning scale economies makes it impossible to conclude that USPS should have a national monopoly over *hybrid* electronic services. Such a determination can be made when sufficient data are available to derive the actual cost characteristics of the emerging industry. Without such a determination, economic theory suggests that this industry might be developed under competitive conditions, although monopoly might be granted in some regional markets until demand evolves enough to support competition.

The second issue of concern is whether USPS should be allowed to offer electronic services. The Postal Service might justify its intention to offer electronic hybrid services for three economic reasons:

1. The USPS system design is eventually intended to serve 95 percent of the national population. No other organizations have proposed a system with such broad public access at such low user costs. This exclusive design stresses a potentially high-capacity system offering nationwide coverage. If Congress agrees with the USPS that a ubiquitous electronic service is in the national interest and no private sector firms step forward to offer services, government may again take the initiative, as it did 200 years ago in conventional mail services.

2. A second justification for encouraging USPS participation pertains to a need for additional operational efficiency within the organization. The claim has been made that "USPS is overly labor-intensive, technologically obsolete in methods of production, and organizationally incapable of developing or evaluating meaningful new innovations."[12] To respond to such criticisms, USPS has attempted to develop innovations that could partially automate its labor-intensive operation. Such efforts, for instance, have led to the development of electronic (OCR) letter-sorting equipment and installation of mechanical sorting machines. The electronic system represents a quantum jump in efforts to automate the postal service.

USPS has maintained an active interest in developing electrical message-transfer capabilities for quite some time. But several of these innovative efforts have been affected by political pressures. In the 1840s the Post Office sponsored many of the early telegraph experiments of Samuel Morse. When Morse was unable to find private investors, the Post Office initiated a telegraph service in 1843. This public service was curtailed in 1847 when Congress refused to renew appropriations.[13] Subsequent attempts by succeeding postmasters general to establish a government-owned telegraph syustem were wholly unsuccessful.[14] Private interests were well organized against any government attempt to assume control of electrical communication services.

During the late 1950s, the Post Office began developing the first precursor of its present design for electronic service. The Speed Mail experiment was a plan to develop a nationwide facsimile network involving about 70 stations. Letters were to be conventionally collected, processed, and delivered and electronically transferred. Between 1958 and 1960, terminals were indeed installed in Washington, Chicago, and Battle Creek, Michigan. But after strong political pressure by the private sector the project was discontinued in 1961.[15] In 1970, USPS operated a facsimile trial service that was curtailed after several months.

Presently, USPS in cooperation with Western Union, is operating two electronic services, Mailgram and ECOM. Mailgram volume has been growing dramatically every year since its introduction. Eventually, Western Union expects Mailgram to revitalize interest in telegraph-like services.[16] ECOM services are just beginning in 1982.

The Postal Service is subject to the same economic realities as other organizations. In a world shaped by technological change, successful firms freely develop and adopt innovations, introduce new goods and services, improve service quality, and reduce consumer cost. USPS is already involved in the provision of electronic services. The adoption of the electronic hybrid technology can be viewed as a logical step to ensure its survival. If political considerations prevent USPS from using such technology, it may be unduly limited to using antiquated postal techniques, clearly an economically inefficient result.

3. Government participation could be rationalized on the grounds that it could act as a stabilizing force in an uncertain industry. Introduction of a USPS system with fixed technical standards could create a calmer industrial climate that would encourage equipment suppliers and service operators to invest their efforts in the unsure industry, especially if potential entrants can be reasonably sure that a competitive environment will result.

Conditions Defining USPS Participation

Some economic reasons have been cited that may justify USPS entry into a competitive electronic hybrid-service market. Because USPS is in a potentially powerful position with respect to this market, some conditions of participation may be necessary if a competitive atmosphere is to be established.

USPS would be clearly interested in offering hybrid electronic services at low cost so that it can develop an adequate market share to justify a large investment. But, if competition is desired, care must be taken to ensure that the service price is not artificially low because of cross-subsidization. Because USPS monopolizes conventional message transfer, prices in that market could be raised to a level that earnings could support electronic service costs. If such cross-subsidization occurs and prices are set below marginal costs, then firms entering this market may not be able to compete with USPS effectively.

The USPS design concept proposes to serve 95 percent of the national population. This scope of coverage can only be achieved in the near term by implementing a hybrid sysem that uses both electronic and conventional methods to convey messages from point to point because of terminal costs. If a competitive market is desired under these assumptions, private firms offering such public services must be able to deliver or arrange for messages to be physically delivered to the proper recipient.

Two alternatives might be considered. First, the Private Express Statutes (PES) might be suspended for messages conveyed electronically, thus allowing private firms to deliver their own messages. There is precedent for such action. Telegrams do not fall under the scope of PES. Second, USPS could be required to interconnect and coordinate competitive electronic-trunk services with the conventional postal-carrier system. Implementation of either of these two measures would facilitate the development of a competitive market.

Thus far the discussion has dealt chiefly with economic considerations relating to the provision of electronic message transfer. However, political, social, and other pressures often inject noneconomic criteria that are somewhat different than the economist's point of view. For instance, value is often placed on dimensions such as political and economic stability, equity, and continuity. Such considerations are no less important than the economic measures.

Social Issues

Hybrid electronic services can be developed either for a selective audience or for universal service. Without government intervention, services may be developed in only the most profitable areas. An inequitable distribution of service might cause those without initial access to harbor resentment concerning a lack of available service. Government may guarantee equity by providing electronic services in less well-services areas. USPS might continue hybrid electronic service development for this reason.

Labor interests within USPS are plainly interested in having USPS-operated services. The National Rural Letter Carriers Association and others have openly urged USPS to continue its efforts for several reasons. First, USPS is unable to make rapid adjustments in its labor pool because of predictable political pressures. Second, labor may have a powerful collective bargaining position because it is dealing with the government as an employer. Finally, USPS might be more sympathetic to the retraining and readjustment needs of workers than would private-sector firms entering the market. It is not surprising that several unions favor a USPS role in electronic services.

The public's traditional notions concerning message transfer have been shaped by public offerings. It has been reasonably confident that the mail service is reliable and that the sanctity of the sealed message will be preserved. This implicit faith, exercised every time a letter is mailed, may be transferred to a public electronic service. Initial suspicion about the new technology may be reduced if government promises to ensure that performance standards are met. Eventually, consumers may view hybrid electronic services as useful alternatives to conventional services. And consumer acceptance of electronic services is crucial to the success of any venture, whether public or private. Therefore it may be desirable for USPS to introduce electronic services to promote public confidence in the new technology.

Government may view electronic record communications services as necessary and essential infrastructure elements much like telephone and conventional mail services. It may be wise to consider maintaining both conventional and electronic services. The addition of electronic services could allow business and the general public flexibility in choosing the most suitable alternative for transmission, and the nation would not be solely dependent upon a single record communication system. Government may, therefore, wish to ensure that such services are developed fully by taking the initiative to build a ubiquitous hybrid system.

Political and Legal Issues

The hybrid electronic service proposed by USPS offers the potential for jurisdictional and legal entanglements. But few would dispute the fact that USPS has the legal authority to develop such a system. One intent of the 1970 reorganization effort was clearly to encourage USPS to seek new methods of increasing organizational efficiency. After studying the technology, a national commission concluded that only end-to-end services would "raise difficult and complex national policy issues concerning competition and regulation."[17] USPS believes that "as long as an electronic message system has physical delivery as the final step, data put through the system would meet the definition of a 'letter.' "[18] The implication is clear that USPS can legally offer a hybrid service.

Such a move might break a longstanding organizational tradition of not competing with the private sector and could be politically unpopular with champions of private enterprise. But because no firms have yet declared an intention to enter the hybrid market on such a scale, this tradition has not been broken. In the event that other firms do enter the market, USPS participation is likely to be hotly contested in the political arena. A recent study suggested that competitive participation by USPS will involve "potentially high political and judicial costs."[19]

USPS may enter the hybrid electronic market and attempt to dominate it. A key issue is whether USPS has the legal authority and desire to extend its monopoly on letter transfers to include hybrid electronic communications. Historical trends reveal such possibilities.

The Postal Service over the years has asserted its authority to expand the basic PES provisions and to adopt enabling regulations. In particular, USPS has repeatedly redefined and expanded its definition of a letter. In 1872, Congress statutorily defined the term *letter* as ''correspondence, wholly or partially in writing.'' In 1909, the Post Office legal counsel speculated that letters had to ''partake of the nature of personal correspondence.'' Almost 70 years later, a letter has been interpreted as ''a message directed to a specific person or address and recorded in or on a tangible object.'' Currently, USPS is arguing for a more vague definition—that is, all ''matter properly transmittable in the United States mail, except newspapers, pamphlets, magazines, and periodicals.''

Over the past 100 years, USPS has been quite successful in extending its letter monopoly as the primary character of the mail has shifted from correspondence to transactions. This trend has led to suggestions that the Postal Service could conceivably halt the private-sector development of EFT by exercising its authority to elaborate on the private express laws.[20] Others have speculated that extending the postal monopoly to include electronic delivery would probably withstand judicial scrutiny.[21] But any such move by USPS would undoubtedly create much political debate.

The Postal Service skillfully addressed the question of competition and monopoly over electronic services in a 1973 report by the Board of Governors. This statement outlined three positions held by USPS over future ventures involving electronic services: first, the Postal Service may compete with private enterprise in this area; second, USPS will not attempt to apply PES to the electronic transfer portions of privately provided services; and third, USPS will apply PES restrictions to output physically delivered output. In other words, although USPS will not extend its monopoly power to include privately provided end-to-end services, it will apply PES restrictions to physical delivery of such messages. This application of PES could put USPS in an advantageous position in a competitive market for hybrid services.

The FCC has recognized this possibility. In the spirit of its desire to stimulate competition in telecommunication markets, FCC has recently challenged the Postal Service's authority to regulate physical delivery of electronic messages. A recent Notice of Inquiry asserts that hybrid services such as Mailgram and ECOM are completely under FCC jurisdiction if it desires to exercise control.[22] This position is based on a reading of the intent of the Communications Act. In a letter to the Postal Service, FCC stated its position regarding development and regulation of hybrid electronic services more strongly:

> Nowhere in the Communications Act is there a suggestion that the development and cost of electronic communications services might be

> subordinated in such a manner to decisions of the US Postal Service, or that such electronic communications Services [sic] might be arbitrarily taxed to subsidize Postal Services. To the contrary, the Communications Act clearly shows that Congress intended that electronic communications services develop as a competitive alternative to postal services and that such services be free and independent of Postal Service regulations.[23]

The emergence of the hybrid electronic technology has clearly brought two traditionally separate statutory mandates into conflict. In addition, PES and the Communications Act have received much recent congressional attention and are themselves subject to revision. It may be in the national interest for Congress itself to resolve this conflict.

Universal versus Selective Service

National policy must be formulated that deals with the question of equity of access. Policymakers may feel that widely deployed electronic services would bind a nation together and facilitate commerce, the same argument used to justify rapid wide-scale development of conventional services. If Congress determines that development of universal electronic service is in the public interest, government necessarily has a role in guaranteeing that this obligation is met. If private developers do not provide services in unprofitable areas, government may find it more efficient to have USPS offer the service rather than to regulate rates and compel private firms to enter these markets. The development of rural electrification may serve as a possible model of government action.

Viability of USPS

The most profitable segments of the electronic market (for example, intraurban transactions mail) will also tend to be the most profitable conventional markets. Diversion of postal patrons into this alternative service may effectively injure the remaining consumers of conventional services who face price increases to cover rising costs. Furthermore, such developments may affect the economic viability of conventional operations. In the long run, if electronic services become quite popular and become a meaningful substitute for conventional service, a crisis similar to that confronted with the railroad industry may emerge. Political decisions may again have to be made but the viability of the conventional system if USPS is not allowed to automate and diversify its services.

Convergence

In the more distant future, our society may be faced with a convergence of the point-to-point and the hybrid-transfer industries—for example, when terminal costs fall to a level where they can be easily afforded by consumers or when consumers discover enough uses to justify acquisition. At that time the Postal Service's role in providing services will again be in question. It would be difficult at this point to declare what that role should be. If competitive market forces provide the incentives for firms to supply all the services demanded at reasonable cost, USPS may find itself playing an unnecessary role. If, on the other hand, public demands are not fully met, continued USPS participation may be appropriate.

Summary

There is little doubt that this country has a preference for rapid written-record transmission services. An important issue of public policy concern is whether electronic message services should be publicly or privately provided and under what regulatory scheme. Two types of electronic services have been distinguished according to system designs that suggest there are two different markets. The end-to-end, wholly electronic services are clearly under the jurisdiction of the FCC and are likely to evolve under competitive market conditions if the current regulatory philosophy continues. The issue of who is an appropriate electronic hybrid services provider is not as straightforward.

The development of the conventional message-transfer service was necessarily initiated by government action. Congress chose to continue a policy of exclusive public ownership to generate sufficient profits to subsidize service expansion because the development of universal service was deemed to be in the public interest. In the case of electronic hybrid services, the conditions are not the same and these justifications may not apply. The private sector now has capital and organizational resources at its disposal. It is clearly able to participate in the development of electronic services if it so decides.

The nation already has a strong, universal record communication system in place. During Colonial times, the development of a message-transfer system was considered essential for stability and growth. Because a conventional service now exists, the need may not be perceived to be great. Alternatively, a nationwide electronic system might be viewed as a new primary infrastructure element worth developing. The government must decide whether selective or universal development of a national electronic system is in the public interest.

A recent regulatory trend in the telecommunication industry has been to encourage competitive entry. The Carterfone and MCI decisions have resulted in the development of innovative terminal equipment and lower service costs for some users. For this reason, FCC and others would like to see electronic hybrid services develop as a competitive industry. Because the cost characteristics of this industry are not yet known, one cannot judge the appropriateness of this approach based solely on economic criteria. USPS wishes to develop such services but in the past has followed a traditional policy of noncompetition with the private sector. The question of USPS participation in an emerging end-to-end competitive industry is another policy issue that may be reexamined at a later date.

A USPS competitive service might be justified for several reasons. Government participation might stimulate market development and inspire consumer confidence in the new services. It might stabilize a cautious industry climate and may reassure the general public that no economic or geographic bias would influence service development. Finally, participation could affect the viability of USPS because the introduction of electronic service could automate a labor-intensive operation and allow new features to be offered.

The competitive entry of USPS into an electronic hybrid market would not be a trivial undertaking. If competition is desired, cross-subsidization of electronic services by conventional services would have to be prevented. Furthermore, physical delivery of electronic messages by private firms would have to be allowed or USPS must be required to offer interconnection with its conventional delivery service at cost. These policy alternatives require careful examination by Congress and the appropriate regulatory bodies.

Another issue that deserves congressional attention is the role of FCC and USPS in regulating these new services. The Communications Act and the Private Express Statutes have been difficult to enforce during the rapid technological changes over the last few years. FCC, the Postal Rate Commission, and USPS have already battled over regulation of a new joint USPS–Western Union offering, ECOM. There are indications that these conflicts will escalate in the future.

Over the medium term, it might be wise to consider having USPS engage in joint ventures with the private sector. This alternative might ultimately be the most economically and policitally expedient course of action. Cooperation with industry may solidify USPS participation in the development of an electronic hybrid service, minimize political and judicial costs, and help to promote equal access to services by the general public.

Notes

1. RCA Government Communications Systems, *Electronic Message Service—System Definition and Evaluation* (Washington, D.C.: NTIS, 1978), executive summary, p. 9.

2. Sèe ''The Postal Reorganization Act: A Case Study of Regulated Industry Reform,'' *Virginia Law Review* 5 (1972), pp. 1030–1098 for a discussion of the Reorganization Act.

3. U.S. Senate, *Evaluation of the Report of the Commission on Postal Service* (Washington, D.C.: U.S. GPO, 1977), pp. 520–523.

4. U.S. House of Representatives. *Recommendations of the Commission on Postal Service* (Washington, D.C.: U.S. GPO, 1977), p. 5.

5. C.E. Ferguson, *Microeconomic Theory* (Homewood, Ill.: Irwin, 1973), p. 498.

6. Morton S. Baratz, *The Economics of the Postal Service* (Washington, D.C.: Public Affairs Press, 1962), p. 67.

7. Rodney E. Stevenson, ''The Pricing of Postal Services,'' in *New Dimensions in Public Utility Pricing,* ed. Harry M. Trebling (East Lansing, Mich.: MSU Public Utilities Studies, 1976), p. 438.

8. Ibid.; J. Haldi, and J.F. Johnston, *Postal Monopoly: An Assessment of the Private Express Statutes* (Washington, D.C.: American Enterprise Institute for Public Policy Research, 1974); ''The Postal Reorganization Act, *Virginia Law Reveiw,* p. 1070.

9. L. Merewitz, ''Costs and Returns to Scale in U.S. Post Offices,'' *Journal of the American Statistical Association* 66: 335 (September 1971), 504–509.

10. U.S. Department of Justice, *Changing the Private Express Laws* (Washington, D.C.: U.S. GPO, 1977), p. 8.

11. Arthur D. Hall, *The Economies of Scale in Today's Telecommunications Systems* (New York: IEEE Press, September 1973), pp. 10–13.

12. Stevenson, ''The Pricing of Postal Services,'' p. 448.

13. R.W. Mayo and W. W. Wittman, *The Structure, Conduct, and Performance of the United States Telecommunications Industry* (Washington, D.C.: NTIS, 1977), p. 22.

14. Wayne Fuller, *The American Mail* (Chicago: University of Chicago Press, 1972), pp. 178–181.

15. Husain M. Mustafa, *The Mechanization and Automation of the*

United States Post Office (Washington, D.C.: Center for Technology and Administration 1964), p. 116.

16. Marianne Karydes, et al., *An Analysis of Domestic Public Message Telegraph Service* (Washington, D.C.: NTIS, 1973), p. 75.

17. National Research Council, *Electronic Message Systems for the U.S. Postal Service* (Washington, D.C.: NTIS, 1976), p. 8.

18. U.S. House of Representatives. *Postal Research and Development* (Washington, D.C.: U.S. GPO, 1978), p. 17.

19. Commission on Postal Service, *Report* (Washington, D.C.: U.S. GPO, 1977), vol. 2, p. 865.

20. U.S. Department of Justice, *Changing the Private Express Laws,* p. 23.

21. Kalba Bowen Associates, *Electronic Message Systems: The Technological Market and Regulatory Prospects* (submitted to the FCC April 1978), p. 179.

22. FCC Docket 79–6, February 2, 1979.

23. FCC letter to General Counsel of USPS, March 12, 1979.

Appendix 4A Public Postal Services in the United States

Historical Origins

The need for conveyance of messages has been recognized in this country since Colonial times. During the mid-seventeenth century, several colonies attempted to organize the sporadic efforts of individuals into colony-wide posts. These attempts failed largely because of poor roads. In 1692 a private monopoly on all colonial postal activities was granted under royal patent. When this service failed in 1707 the British government assumed control. The original private monopoly of 1692 was the only instance in which the letter transfer function was ever privately provided in the United States.[1]

During the Revolution, one of the first concerted actions against the crown was the founding of a postal service that linked all of the colonies and also ensured that traitorous messages would not be intercepted or read by British authorities. A dependable postal system was necessary both for the exchange of military information and for encouraging the rebellious spirit of the citizenry. The value of such services to the nation's founders is evinced by the explicit attention given to postal activities in both the Articles of Confederation and the Constitution. More important, in each case the provision of postal services was defined as a public undertaking.

The U.S. government, as did governments in many other countries, historically took the initiative to provide postal services. Moreover, such services have continued to be publicly provided to achieve a number of objectives. There were two reasons for establishing a message system that was outside of the control of the British authorities. The normal expectation in that day, and indeed even in the early days of the national postal system, was that mail was often opened and inspected. Furthermore, mail often did not reach its destination. The young government was clearly interested in establishing a reliable system for the security of the state, beyond the control of any individual entrepreneur.

Second, the young government desired to unify and bind the colonies (and later states) together before and after the war. Without a strong and dependable communication system, the infant country risked the emergence of factionalism and regionalism. Fearing this development, Washington allocated precious funds from the small national treasury for the development of a postal system.

The third and related objective was the desire to develop and centralize the political power of the national government. After the Revolution, the postal system was used to propagandize as well as to serve as a very visible example of services provided by the national government. Later, as the nation expanded westward, citizens of outer territories were drawn toward statehood by the promise of roads, postal services, and other federally provided enticements.

A fourth national concern was to build solid components of a national infrastructure that would ensure the development of commerce and cooperation among the states. The importance of transportation and communication linkages among states was clearly implied by the explicit constitutional direction to the federal government, rather than the states, to provide postal services over post roads.

All of these objectives could not have been met by one or more entrepreneurs without the initiative of government. Much of the task of creating a reliable postal system initially depended on an ability to build and maintain usable roads. In fact, many of the nation's roads were constructed under the constitutional mandate to furnish post roads. Few organizations had the resources to undertake such construction. Hence, the federal government necessarily took the initiative to provide services.

Challenges of the Private Sector

Thus far the discussion provides a partial explanation of the origins of public provision of postal services. It does not, however, suggest a rationale for continued adherence to this doctrine. This section highlights the challenges of the private sector and the political response to such attempts.

During the nineteenth century much of the country's infrastructure was in the midst of rapid development. Post roads continued to increase at a rapid pace. Post roads extended from almost 2,000 miles in 1790 to more than 65,000 miles in 1820, all financed by postal revenues. At the same time, various modes of transportation services were also developing. Private stage lines carried passengers and packages throughout the states and into the frontier on these roads. In the 1830s, a very young railroad industry was being nurtured by government land grants and liberal tax policies. All these transportation modes were hired under government contract to carry the mail. These developments provided the necessary incentives for the emergence of serious challenges from the private sector.

Many transfer companies used these infrastructure developments to facilitate the flow of commerce and travel throughout the expanding borders of the country. Express companies gathered parcels in cities and towns and delivered them to designated destinations. By the 1820s, the private express companies had also developed a solid reputation for rapid and secure delivery of letters, packets, and other articles.

The public responded enthusiastically to this cheap and reliable message-transfer service. It is estimated that by the 1840s, one-third of all documents were being carried by private companies. The famous pony express, a characteristic feature of our country's westward expansion, was one such private undertaking. By 1844 the private-sector challenge reached a point where the Wells and Company messenger firm proposed to take over postal activities and boasted that a uniform postage charge of 5 cents was possible.

Congress did not ignore the competitive response of the private sector. Throughout the first 50 years of its existence, Congress enacted laws to strengthen the ineffective government monopoly on letter mail services. As early as 1791, carriage and packet-boat companies were prohibited from offering private mail services. Between 1827 and 1838, prohibitions were extended to foot and horse posts, canal boats, and railroads. Yet the number of letters carried outside of the government service continued to increase.

Finally, in the 1840s, the Post Office Department asserted its legal authority. In 1840 and 1843, the Post Office sued two private companies and lost both cases. The courts' actions essentially rendered the postal monopoly unenforceable.

Congress, recognizing the Post Office's plight, responded by passing the 1845 act, which lowered the postage rate and prohibited private express carriage of letter mail. The congressional debate over this act has been characterized as the only major serious congressional consideration of the postal monopoly. The hearings reveal four prevailing thoughts. Congress believed that the post office could not hope to compete with private express companies; that government had a duty to provide services to nonpaying frontier and rural areas and the services should continue to expand; that losses on the frontier should be subsidized by the profitable services; and that legal monopoly was necessary for cross-subsidization.

The 1845 act effectively reinforced the previously uncertain monopoly. The reasons for enactment are fairly clear. A reliable source of revenue was badly needed to finance both service improvements and service expansion. The monopoly was intended as a revenue protection device. Throughout the history of the government postal monopoly, these revenues have been used to expand the system of post roads, increase the number of post offices, and provide both city and rural delivery systems.

After 1845, the Post Office monopoly on letter mail did not receive a great deal of major congressional attention. Between 1945 and 1970 the statutes were recodified on several occasions. Today, they are referred to as the Private Express Statutes. The only significant congressional event occurred in 1934 when Congress, again fearing erosion of revenues from the activities of Western Union messengers, limited special messenger service to 25 letters per client. Since then, the Postal Service has continued to expand its monopoly by means of its authority to interpret the meaning of ''letters'' in the statutes broadly and by actively enforcing existing laws.

Notes

1. See Wayne Fuller, *The American Mail* (Chicago: University of Chicago Press, 1972), and George Priest, ''The History of the Postal Monopoly in the United States,'' *Journal of Law and Economics* (April 1975), pp. 33–80 for additional discussion of the postal monopoly.

5 The Impact on USPS and Resource Considerations

Knowledgeable sources both inside and outside of government expect that at least some of the messages carried over both end-to-end and hybrid electronic message services will be at least partially diverted from the conventional postal letter stream. Projections in chapter 3 indicated that between 11 billion and 35 billion first-class items could be transferred over electronic systems if business users and the general public respond favorably to the new technology. With mail volume continuing to increase at current rates, such a diversion could represent as much as 50 percent of future conventional message traffic.

Over the next decade, USPS hopes to develop a mature hybrid electronic system that will partially automate the handling, sorting, and transfer of up to 25 billion first- and third-class mail items. If USPS is permitted to continue developing the hybrid system, it will be well positioned to offer electronic services in the event that they become popular. If USPS is unable to develop electronic services, it may experience a decreased market share and suffer major cuts in its revenue base. USPS may eventually face such choices as radically changing productive capacity, putting greater reliance on much larger subsidies, or dramatically raising rates. A rate increase could make the price of conventional services much higher than the price of electronic services, perhaps encouraging an even larger diversion of messages to the cheaper services. Thus, an increasing popularity of electronic services will have a significant impact on USPS, regardless of whether it participates in service development.

USPS could face several structural changes in its attempt to alter its production process. If USPS offers bimodal electronic and conventional services, it will be necessary to make a substantial investment of capital and to transform the role of its labor force in service operations. If USPS provides only conventional services in the future, it might streamline productive capacity or reduce service frequency (for example, by reducing deliveries to three days per week) to adjust to the diversion. In any event, without a change in rates or an additional subsidy, diversion of letters to electronic systems will have a significant impact on USPS operations regardless of whether it develops electronic services. To appreciate the magnitude of this potential impact it is necessary to known something about the conventional postal production process as well as its cost distribution and revenue-generation procedures.

Conventional Postal Operations

In 1980, USPS delivered more than 106 billion items; of these, almost 64 billion were first-class messages. This task required an average of more than 40 million stops per day, amounting to the equivalent of four round trips to the moon for rural routes and 1 million miles for city routes.[1] As one might expect, USPS operations represent a unique assembly of resources and management practices.

Probably the most outstanding characteristic of USPS is its massive size and ubiquitous nature. It is one of the largest corporations in the world, established by constitutional mandate. If it were rated in the Fortune 500, it would be among the top ten. It employs almost 1 percent of the total U.S. working population, or about 667,000 employees, most of whom are unionized. USPS controls an immense amount of real estate, directly owning 3,000 facilities and leasing 26,000 more for a total of 5 square miles of plant; it maintains about 30,325 post offices and 9,160 branches and stations to provide public access to the system. USPS has one of the largest vehicle fleets in the world. It currently owns approximately 120,000 vehicles in active service, with another 70,000 under contract. This massive transportation pool is larger than the combined fleets of the top five commercial carriers. The physical plant, vehicle fleet, and labor force managed by USPS far exceeds that available to the very largest corporations such as AT&T.

The general public is largely unaware of the internal workings of the Postal Service.[2] More knowledgeable observers find it remarkable that USPS is able to accomplish the massive job of managing the production process and coordinating daily operations, given the uncertainties in demand and the instability of input factor costs (including capital, energy, and labor). Yet increasing numbers of citizens are aware of the rising deficit resulting from current operations. Emerging financial problems have prompted a wide range of suggestions such as increasing mechanization of postal operations or encouraging private-sector provision of postal services. Others have proposed that electronic services be incorporated with conventional operations. Each proposal would have only a limited impact upon the deficit problems.

The postal system has various features that comprise its cost structure. Historically, postal operations in every country have always been labor-intensive activities. Since the Reorganization Act of 1970, USPS postal salaries and benefits have been indexed to similar positions in the private sector. As a result, labor costs now make up 85 percent of the annual postal budget (the budget for the 1980 fiscal year was $19 billion). Recent union agreements that provide job protection and salary adjustments for inflation guarantee that labor costs will continue to make up a very high proportion of the operating budget over the near term.

USPS has attributed its financial situation to the quality of service provided. Fifty percent of the costs are considered fixed, institutional costs associated with making the service available. The other 50 percent depends on the volume of mail carried. This division represents an extremely high fixed-to-variable-cost ratio. Moreover, union agreements limit the employment of part-time and temporary personnel, requiring USPS to carry employees usually needed primarily during peak periods (such as before Christmas) for the entire year.[3]

Operational costs can be disaggregated further. The processing stage has consistently made up one-third of total costs. Transportation costs between postal facilities have fluctuated between 10 and 15 percent of the budget, whereas collection, acceptance, and delivery account for most of the remaining costs.[4] The high proportion of expenditures associated with collection and delivery can be attributed to the labor costs of carriers, which depend on the number of delivery points served and frequency of delivery rather than on the volume of mail carried. As a result, carrier costs and, correspondingly, total costs can be expected to continue to rise if the increasing number of delivery points are serviced according to current practices.

Finally, general economic conditions have also had a large impact upon the USPS budget. For example, rising energy costs have a considerable effect on the USPS operating budget. In 1977, the $.30 increase in the price of gasoline necessitated a $105 million hike in the budget. The escalation of petroleum prices is likely to have a substantial future impact on the total budget. In addition, because so much of the annual budget goes to salaries and benefits, each $250 salary increase per workers implies a total budget increase of $150 million. All these factors contribute to a fairly predictable annual increase in total expenditures.

Postal expenses are met through a variety of means. The primary source of income is derived from postal charges for items carried. Revenues have seldom exceeded expenses in the era of modern postal operations. During the last decade, annual revenues typically accounted for 80 to 85 percent of yearly expenses. In fact, since 1950, there have been revenue surpluses in only 9 years, despite the rapid rise of postal rates. Before 1972, the government appropriated funds to cover each yearly shortfall. As a result of the postal reorganization, USPS was directed to raise more revenue to cover expenses completely.

A related consequence of the reorganization was that postal appropriations became more tightly controlled and strictly defined. USPS has received an annual subsidy for "public service" to rural areas, which will drop from $920 million in 1972 to zero in 1984. USPS may request $460 million for subsequent years. In addition, the government provides compensation for revenues lost from franking privileges and mailing discounts (for example, reduced rates for nonprofit and publishing firms) mandated by legislation. In 1981 this compensation totaled almost $800 million. New legislation will reduce this appropriation by approximately $100 million in subsequent years.

The revenue-generation process has several unique features. Rates set by the Postal Rate Commission have been structured such that first- and third-class items cross-subsidize other classes of mail. Based on this rate structure, first-class mail generates about 58 percent of revenue collected, with transaction mail providing four-fifths of that business.[5] The development of cheaper alternatives to USPS service, either conventional or electronic, would probably force a change in cross-subsidization practices and possibly reduce overall revenues. These revenue-protection arguments have been used to rationalize strict PES enforcement for over a century.

In addition, revenue is generated by only a small fraction of all postal facilities. Approximately 6,000 post offices collect over 85 percent of all revenues. USPS and GAO have argued that service standards could be met at a lower total cost if some of the lowest revenue-producing facilities were closed. But the public has consistently argued for retention of these facilities on the basis that the network of 40,000 community post offices is a public good.

Over the last decade, labor costs have been reduced by introducing more mechanization and automation into postal operations. Yet annual shortfalls have continued to mount, despite rate increases for all mail classes. Within four years after the postal reorganization, a deficit of almost $2.8 billion had been generated.[6] This deficit was not the result of any sudden increases in organization inefficiency. Despite increases in mail volume, USPS has reduced labor requirements by 53,000 work years since 1971. The negative balance accumulated since the reorganization may be attributed to a change in philosophy about who should bear the costs of postal operations.

The postal system is a hierarchically structured operation. The central core consists of about 550 processing centers, each serving a collection of post offices (figure 5–1). These centers have been established in major cities and in state districts. All mail traveling between post offices passes through these area centers. Only local mail is directly processed at home post offices. Centers that exchange high volumes of mail are directly linked (for example, Chicago and Los Angeles). The lower-volume or-

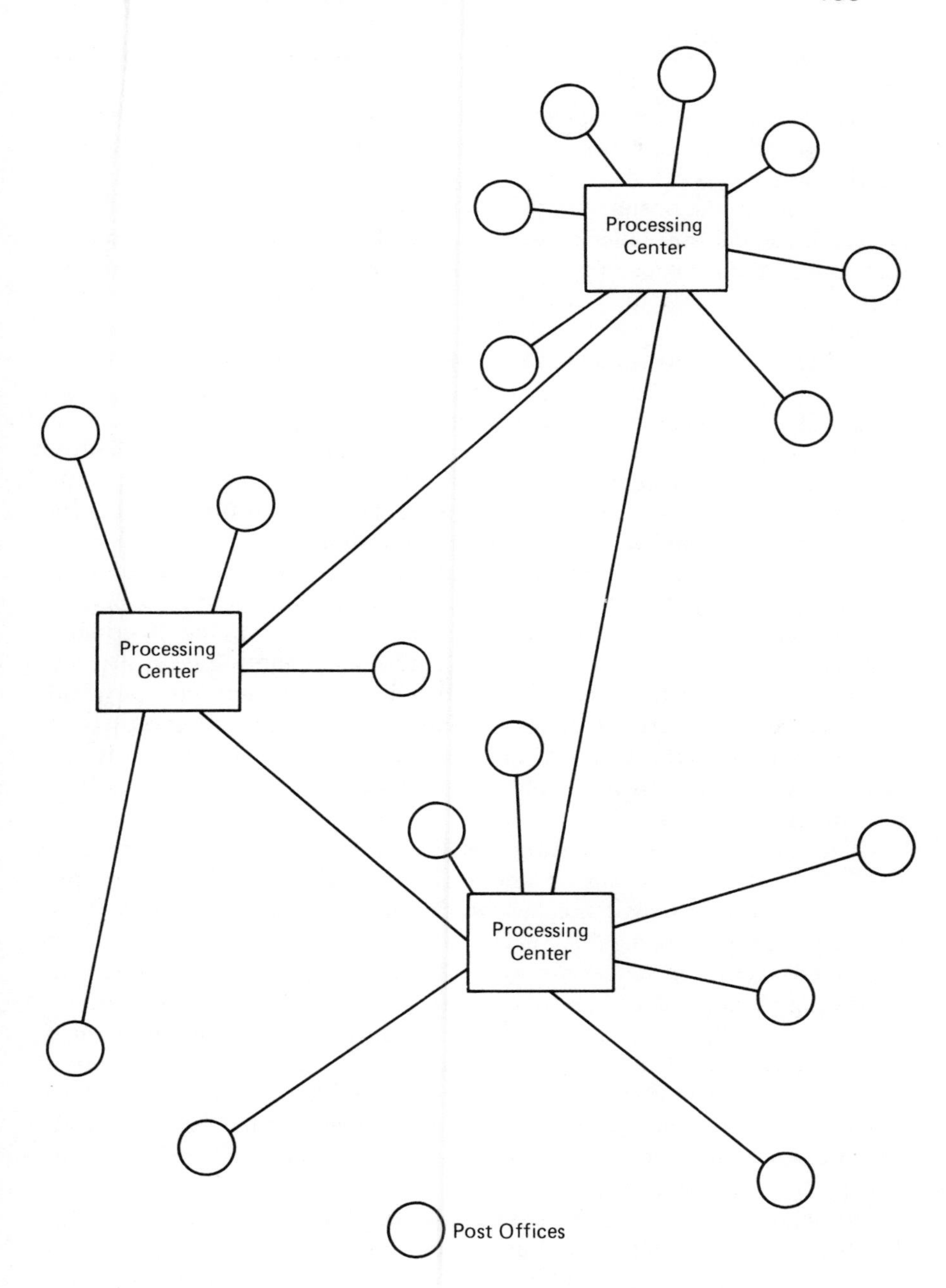

Figure 5–1. Mail-Processing System

igin-destination pairs are linked in a sequential fashion for reasons of economy.

These facilities are used to accomplish four principal processes in postal operations: collection and acceptance, processing, transport, and distribution.

Mail enters the postal system by a variety of means. Daily collections are made at 350,000 letter boxes. Mail is also picked up by carriers as they travel their routes. Finally, items are accepted at post offices by postal staff. Most of the collection activity is carried out by motorized carriers.

Processing involves handling and sorting items at various stages while the mail is en route. Basically, three types of mail require processing. Locally collected mail is composed of local and out-of-town items. Transit mail includes items that are only passing through a particular facility during a movement between origin and destination points. Finally, incoming mail consists of items that have reached their destination facility and must be fed into the local distribution system.

As mail moves through each facility, each item must be handled before and after each sorting operation. These handling activities include loading and unloading of vehicles, movement to and from the processing line, and carriage between operations. Additional handling tasks include opening sacks and bundles, *culling* (that is, sorting items by physical attribute), *facing* (turning items so that their stamped side is up), and *canceling* (postmarking items and invalidating further use of postage stamp). All these steps in handling typically take place at the local processing center.

Currently, numerous sorting steps are required during processing. Collected mail is first sorted into local area and out-of-town groups. When the outgoing mail reaches its destination center, it is sorted by delivery-area post office. At each post office, incoming mail is then separated according to carrier route. Finally, each piece is sequenced according to stops along the route every morning for 1 hour or more by street carriers. The processing effort is not only time-consuming but also very expensive. In 1980, 229,000 clerks and mail handlers were required to process more than 106 billion pieces of mail, representing over 40 percent of the regular postal work force. Mail sorting has become much simpler since the introduction of five-digit zip codes in 1962 and presorting by customers.

Probably the most significant development to date in mail processing has been the large-scale adoption of mechanical sorters. This equipment reduces the amount of handling required in letter processing by automatically canceling and mechanically moving letters into sort bins. The machine operator reads the zip code and enters the proper digits into a keyboard to direct the letters to the proper bin. Mechanical sorters now handle more than 60 percent of all letters sorted.

Mechanization of sorting activities, while requiring a significant investment of capital, has reduced the number of clerks and mail handlers despite an increase in the amount of mail processed. Another consequence of mechanization is that local post offices must now consolidate mail at a central facility to provide enough volume for the sorting equipment. In some cases, items travel much farther than they would without consolidation. Also, mail may be delayed while it awaits the arrival of enough volume to justify machinery operation. With mechanization, error rates in the sorting process tend to be much higher than by manual methods.[8] Misrouted mail is the primary cause of late delivery.

Mail moving between post offices depends on two basic transportation systems. The intracity network that connects offices and substations with processing centers is operated by USPS. Postal Service employees in USPS vehicles typically make two daily runs in each direction to deliver incoming mail to offices and carry outgoing mail to processing facilities. In addition, drivers shuttle mail between facilities and intercity network terminals. Such terminals include rail depots, airports, and trucking facilities.

The intracity transportation system uses a diverse mix of vehicles ranging from small trucks to tractor-trailer combinations. In 1978 about 12,000 vehicles were used to shuttle mail among facilities. The remaining 108,000 postal-owned vehicles are used in the delivery operation. Approximately 11,000 people are required to operate the shuttle system, 55 percent of whom are drivers while the remaining personnel provide vehicle service.

The long-haul intercity transportation network operating between processing centers is a multimodal freight operation. USPS contracts for services with railroads, trucking firms, and airlines for 4-year periods. These contractual agreements typically include time schedules for trips, vehicle capacity, origin-destination points, and rights to purchase extra trips at specified rates when heavy volumes need to be moved. Table 5–1 illustrates how costs of the intercity network have risen over two decades in response to increasing volume.

Although postal operations have used highway, rail, and air transportation since each became available, the mix of modes has shifted dramatically over the last two decades. Table 5–2 shows the percent of total payments by mode in selective years. The most significant change has been the dramatic shift from rail to highway and air transport. Between 1955 and 1970, rail payments fell from a position of undisputed dominance to parity with the other modes and it is now well below motor and air modes. This shift is the result of drastic reductions in passenger train

Table 5–1
USPS Transportation Costs
(millions of dollars)

Fiscal Year	Domestic Intercity Air	Domestic Intercity Motor	Domestic Intercity Rail	Foreign[a]	Other	Total[a]
1955	34	40	297	43	35	449
1960	49	53	342	65	47	555
1965	82	100	331	88	29	630
1970	165	185	182	168	45	745
1971	171	198	167	172	41	749
1972	.176	219	124	146	42	707
1973	215	242	105	142	32	736
1974	215	265	110	146	31	767
1975	217	286	117	196	35	851

Source: Revenue and Cost Analysis Division, USPS. ''Summary Description of USPS Development of Costs by Segments,'' July 1977, p. 130.

[a]Includes costs for transport of military mail directly reimbursed by the Department of Defense.

Table 5–2
USPS Transportation Costs
(percent of total dollars)

Fiscal Year	Domestic Intercity Air	Domestic Intercity Motor	Domestic Intercity Rail	Foreign	Other	Total
1955	7	9	66	10	8	100
1960	9	10	61	12	8	100
1965	13	16	52	14	5	100
1970	22	25	24	23	6	100
1971	23	27	22	23	5	100
1972	25	31	17	21	6	100
1973	29	33	14	19	5	100
1974	28	35	14	19	4	100
1975	25	34	14	23	4	100

Source: Revenue and Cost Analysis Division, USPS. ''Summary Description of USPS Development of Costs by Segments,'' July 1977, p. 131.

Note: These shifts in the mix of transporation services purchased continue to occur less for reasons relating to the volume of particular types of mail than for reasons traceable to service availability, technology, and managerial discretion. Mail service standards, convenience in managing the overall flow of mail, and the foreseeable costs associated with available alternatives determine business decisions on purchases of transportation services.

The cost of this segment in FY 1975 was $769.2 million, excluding $81.8 million for military-reimbursement mail.

service, the need to meet delivery performance standards, and for convenience in managing the flow of mail.

Air and bus contract mail generally moves according to passenger schedules. The Federal Aviation Act of 1958 directed airlines to give priority to carrying mail after passengers and baggage were loaded on each flight. Postmasters eventually responded enthusiastically by increasing the volume of air-transported items by over 20 percent in a 1-year period. Although the average air haul is 1,000 miles, nearly all first-class items traveling 300 miles or more have been transported by air.[9]

When airline passenger schedules have not met the needs of postal operations, USPS has used air taxi services. This limited service, which began in 1967, grew essentially because of a declining number of scheduled night flights (the period when sorted mail must be transported to processing centers). Air taxi services, however, are very expensive since USPS contracts for the exclusive use of aircraft to carry small amounts of mail on a round trip basis. As a result, between 1976 and 1979, the number of air taxi routes has decreased from 170 to 50 flights. In the future, air taxis may be phased out because overnight truck transport could accomplish the job efficiently.

Most first-class items that do not move by air are moved by truck. Highway transport is the most flexible mode with respect to points traveled and frequency of travel. Trucks are commonly the most efficient mode for hauls within a 250-mile radius.[10] USPS currently has over 12,000 contracts, principally with small carriers employing about 16,000 persons and leasing 16,000 vehicles. The typical firm consists of one operator with a single vehicle, which is either a 27-foot or a 40-foot tractor trailer. Because mail volume has seasonal and yearly variation and truck capacity is fixed for 4 years by contract, some scheduled trips have unused truck capacity.

Rail service is used for long hauls, primarily for second-and third-class bulk loads. USPS commonly uses trailer-on-flat-car (TOFC) and occasionally runs dedicated mail trains. At one time, postal operations depended on post office rail cars where mail was sorted while being transported. However, most of this service was discontinued in 1971. The last railway car service, between New York and Washington, D.C., was discontinued in 1977.

Mail is distributed by two basic methods, either by carriers on foot or by vehicles owned or leased by USPS. A surprisingly popular alternative is customer pickup from post-office lobby boxes or windows. As much as 40 percent of all mail is not delivered by carriers, implying that the cost of distributing only 60 percent of the mail amounts to roughly 15 percent of the annual budget.[11]

Carriers deliver mail according to three different types of route assignments. City carriers are USPS employees who distribute mail to more than 68 million addresses along high-density urban routes. Rural carriers are USPS employees whose routes are much longer, typically 70 miles each, but with a lower number of delivery points. Very sparsely populated areas or difficult-access sites that do not qualify for rural delivery service are covered by contractors who bid competitively for their jobs. For example, delivery routes for island communities are typically served by contractors who supply their own boats. Rural delivery service reaches more than 14 million addresses.

Although delivery points have been increasing by 2 percent annually, USPS has been able to contain the growth of resources necessary to distribute mail to these points. Nearly all the more than 32,000 rural routes are motorized under vehicle leasing arrangements with carriers. By motorizing 93 percent of its 127,000 urban routes, USPS has been able to curtail the growth in the number of routes. The number of urban carriers has declined by 27 percent to 157,000 persons since 1971. To achieve these productivity levels USPS operates a delivery fleet of about 108,000 vehicles and leases 19,000 more.

Modification of Conventional Operations by Electronic Technology

Postal officials have been studying the possibility of using telecommunications and electronic hardware in mail services for over 20 years. The development of Speed Mail, referred to in chapter 4, was its first serious attempt to incorporate communications technology in conventional operations. Since then, several other services have been planned and implemented. A variety of contractors have studied the feasibility and cost-effectiveness of a large-scale hybrid system like that described in chapter 2. In addition, USPS continues to look toward successful development of OCR sorters, bar-code readers, and other electronic aids.

In the last decade, various combined systems have been proposed. Although the actual designs differ in such areas as message capacity, equipment configuration, number of processing stations, and labor requirements the central idea—the integration of electronic and conventional postal operations—has been fairly consistent. None of the designs have been planned for a capacity greater than one-half of projected future postal message volume. The latest proposal suggests that a well-planned USPS system may be successful even if it does not carry EFT traffic, which might be diverted to specialized systems. The central point is that electronic operations are not being designed to replace conventional services totally or to dominate the structure of current operations, at least in the immediate future. Therefore, both services could coexist with some cost-sharing and integration of production activities.

In addition to its conventional collection activities, USPS could collect messages from 45,000 electronic terminals installed in public areas (for example, airports and shopping malls). Such facilities would accept both individual and bulk packets of messages to be converted to machine-readable form along with encoded messages already stored on magnetic media (such as tapes or cards). Finally, postal stations would accept encoded messages sent over telecommunication lines. These additional collection-acceptance activities would require greater capital investment, more technically trained personnel, and different management procedures.

Processing alone, as designed, would require a very large commitment of new equipment. Current USPS designs call for the development of eighty-seven processing stations at the largest conventional processing centers to handle the electronic message volume. These stations would be equipped with specially designed paper-handling machinery, message-encoding devices, magnetic storage readers, storage facilities, computer-processing hardware, and printing equipment. Operating budgets for this service must include allocations for consumables such as paper and ink because hard-copy message production inherently would be part of the hybrid scheme. Current designs would also result in a reduced staff and a different mix of employees, a subject discussed further in chapter 6. Multiple sorting and handling needs may be drastically reduced for electronic services if individual and bulk users assist in electronically encoding destination addresses and postal codes.

Electronic services would rely heavily upon the intracity transport network. Messages printed at the eighty-seven stations would be moved to local post offices by USPS personnel and vehicles. If the priority services became popular, more frequent and lower-capacity trips would be likely. On the long-haul intercity network, substantial modifications may be necessary. If electronic services divert letter mail from conventional services, truck capacity may be reduced on the shorter runs while airlines may lose revenue from reduced mail volume. Alternatively, trip frequency could be reduced, which may significantly decrease the speed of conventional delivery service.

To transfer electronic messages, USPS would require access to telecommunications equipment that protects the security and integrity of the messages. At each of the 87 proposed stations, such equipment would handle transmission and reception needs, as well as error correction and message ciphering procedures. Increased popularity of electronic services could reduce capital and labor requirements of the conventional intercity long-haul network.

Current plans indicate that electronic messages would be distributed either by delivery along conventional routes or by customer pickup, suggesting that the delivery force may be left intact. Some competitors fear that USPS may eventually operate local telecommunications links for input and output directly between stations and messge origin-destination points. Others feel that provision of all electronic services to "hard-to-reach" rural areas would be a cost-effective move for USPS. Such an expansion of service might reduce both carrier and vehicle requirements. However, suggestions that USPS operate either electronic collection or distribution services have been met with quite vocal opposition.

Electronic Services and USPS

The advent of electronic message transfer will necessarily affect the future of USPS. Two very simple questions can be posed. If USPS is barred from offering electronic services, can it and should it survive? If USPS is allowed to compete with others, will electronic and conventional operations be successful? The question of survival is dealt with first.

If USPS is excluded from the electronic market, its viability will depend on consumer reactions to *both* conventional and privately offered electronic services. Consumers may distrust new electronic services or simply prefer to use conventional services, thus ensuring the continued existence of USPS. Users might view both options as functionally adequate and sufficiently equivalent for some message-transfer needs and select the presumably cheaper electronic service. In that event, USPS might align first-class rates more closely with costs and forego the traditional practice of cross-subsidizing the other mail classes in an attempt to reduce the diversion of messages from its conventional service. Thus, USPS may find itself in a legally competitive market to provide message services for the first time since it was established, despite standing Private Express statutes.

In offering electronic services, USPS could undergo a radical evolution that might prompt such organizational changes as altering the nature of the workplace, shifting the capital-labor mix, changing its traditional public relationship and image, and entering into a new regulatory framework. For instance, USPS would like mailers to assist in electronic sorting operations by having mailers key in message destination postal codes at electronic mailboxes when individual letters are being dropped off. Also, USPS would like to accept responsibility for message production along with transfer services because new technology makes integrated message-production and transfer services more economical than conventional separated practices. Successful operation of this bulk system requires that users become accustomed to submitting messages stored on magnetic media with the confidence that they will be transferred and translated accurately. Therefore, USPS will be asking individual mailers using the electronic services to do a little more and bulk mailers to do less. These actions, which are by no means radical, will require that the public accept a different image of USPS service delivery and also play a different role in service operations.

USPS has become actively involved in marketing since the reorganization in an attempt to overcome its traditional image forged through 200 years of service. This activity has promoted product diversity and service flexibility. In the future, USPS is likely to encourage the public to accept it as a provider of very rapid message delivery service (within 1 or 2 hours) and message-transcription services. Success in the electronic message-transfer field will depend on public confidence in the new services.

Because labor costs in processing represent almost one-third of total operational costs (and those costs now rise automatically because of cost-of-living salary adjustments), it is fairly obvious that substitution of capital for labor could affect the size of the required budget. In a mature system, electronic processing and transmission could cost as little as $.018 per piece compared with $.08 or $.09 for equivalent conventional operations. If bimodal USPS operations evolve to a point where the number of electronic messages is a significant portion of total message volume, USPS may alter its capital-labor ratio substantially as a budget control measure. Such activities could lead to energy savings and reductions in labor requirements, but would require an additional investment of about $1.8 billion for equipment. The new effect of these resource shifts may reduce costs by as much as 40 percent.[12] (Energy-conservation possibilities are considered in the next section; labor issues in chapter 6.)

If USPS successfully deploys and operates a hybrid electronic service, there may be some future incentives to substitute capital for labor in the acceptance, collection, and delivery phases of the electronic service. The public may embrace the new technology by purchasing low-cost home terminals, prompting USPS to take a more competitive and aggressive role in the provision of all-electronic message services. However, this action would not bring about the same scale of reduction as in the processing case, unless a large majority of postal patrons prefer electronic over conventional services.

As long as conventional services remain popular, the workforce and vehicles necessary to service 80 million delivery points will be required. USPS might consider incorporating such diverse services as grocery delivery (as proposed for British Post Office operations) to generate enough revenues to maintain distribution routes in the face of shrinking message volume. But again, this alternative casts USPS into a new role that may not be readily accepted by the public.

The initiation of electronic services gives USPS the opportunity to exert more control over its operations. Although conventional operating procedures are governed by a voluminous general manual, everyday decision making is rather decentralized. The electronic service will permit less autonomy because satellite channel-loading schedules and sorting activities will be centrally coordinated and scheduled over the entire system, perhaps enhancing delivery performance and alleviating peak-load problems. Also, by reducing its commitments to 4-year transportation contractors, USPS may gain more flexibility and control. However, the provision of electronic services may prompt public service commissions and telecommunictions regulators to join CAB, PRC, and ICC in regulating USPS activities.

In the future, USPS might subsidize a faltering conventional service with funds from its booming electronic service, especially if electronic costs are much below conventional ones. But cross-subsidization would be difficult in a vigorous competitive market. If electronic services (USPS or otherwise) divert a substantial message volume away from conventional service, conventional rates may rise or services may be reduced. The problem of a shrinking revenue base might be alleviated by taxing every electronic service to subsidize a highly valued conventional service (for love letters and other personal messages), or conventional service could be supported by general tax revenues. If conventional costs are streamlined by reducing service, a situation may develop in which electronic services are used for rapid transfer of time-sensitive messages and conventional services carry the less routine, more personal communications.

According to one knowledgeable observer, USPS operations strive to meet two fundamental objectives: avoiding delay in message services and reducing unit-handling costs.[13] Electronic services are being designed to address both objectives. Because telecommunications permits transfer times to become independent of communicating distances (that is, message-transfer time is a function of system load and delay may result as messages wait in channel-loading queues), operational delays implicitly associated with the long-haul conventional transportation system could be avoided. Automated electronic sorting will also reduce many sorting requirements, thereby further increasing processing speed.

If USPS successfully implements the proposed hybrid system according to design specifications and if annual volume grows to 25 billion messages, costs per piece of electronic messages could be one-third less than conventional costs. Cost studies suggest that annual savings could exeed $200 million.[14] Future savings could be even greater because investment costs will not require the inflationary wage adjustments of the labor replaced. By adding electronic delivery to the hybrid design, USPS may further reduce total annual operating costs. For instance, some of the $190 million spent on paper and ink may be unnecessary. Hybrid-system printer capacity and conventional-delivery input factors (such as vehicles or carriers) might be partially reduced if some message traffic migrates to an all-electronic transfer service. Such a modification in service would undoubtedly enhance the cost-savings potential of a USPS electronic message-transfer system.

Energy Implications

The energy requirements of conventional postal technology can be compared to those of electronic message transfer. Any complete comparative analysis would require an examination of the total energy used by each system (that is, the energy required to *produce* necessary machinery along with energy used to *operate* equipment). Because such data do not yet exist, such an analysis is impossible at this stage of development. However, the projected energy utilization for electronic transfer operations can be compared to that required for conventional transportation of messages.

Table 5–3 lists the labor, equipment, and power requirements of a model electronic hybrid-transfer system, with power requirements for the four types of transmission station. A complete transfer system, consisting of a varied mix of these stations, would use 25,141,200 kilowatt-hours or 85.78 billion Btus to move 30 billion messages. On a per-message basis, each transmission would require an average of 2.9 Btus.

Table 5–3
Yearly Resource Requirements of an Electronic Message-Transfer Hybrid System[a]

Transponder Lease[b]			$750,000
Power			
Terminal A	50 kilowatts = 438,000 kilowatt hours each/year		
Terminal B	30 kilowatts = 262,800 kilowatt hours each/year		
Terminal C & D	20 kilowatts = 175,200 kilowatt hours each/year		
Staffing — 3 shifts/24 hours			
Terminal A & B	1 Engineer/shift @ $40,000	=	$120,000
	2 Technicians/shift @ $25,000	=	$150,000
Terminal C & D	1 Engineer/shift @ $40,000	=	$120,000
	1 Technician/shift @ $25,000	=	$ 75,000
Staffing — 2 shifts/24 hours			
Terminal A & B	2 Technicians/shift @ $25,000	=	$100,000
Terminal C & D	2 Technicians/shift @ $25,000	=	$100,000
Control — 3 shifts/24 hours			
Terminal A, B, C, D	2 Engineers/shift @ $40,000	=	$120,000
Control — 2 shifts/24 hours			
Terminal A, B, C, D	1 Engineer/shift @ $40,000	=	$ 40,000

Source: Philco-Ford Corporation, *Conversion Subsystem for the Electronic Mail Handling Program* (Washington, D.C., NTIS, 1973), task 4, p. A–24.

[a]Assumes thirty billion messages per year.

[b]Minimum 4, maximum 7

To estimate the energy utilized by the equivalent conventional method of transportation, several assumptions must be made. First, the transportation of messages between conventional processing stations and transportation terminals (now provided largely by the USPS fleet) is assumed to be similar to the transportation requirements of moving hard-copy messages among the electronic-processing centers and the conventional distribution centers. This simplifying assumption suggests that the transportation requirements of the intercity contract fleet will be equivalent to those of the electronic transfer activity. A second necessary assumption involves the distribution of messages carried by each mode. Nearly all interurban first-class messages travel on airline passenger flights; the remaining messages move by truck. Because less than 60 percent of first-class mail is local, in this sample calculation 40 percent of the message load is assumed to move by air while the remaining 60 percent is moved by truck.

In 1977, USPS estimated that 400 million ton-miles of transportation moved 53.7 billion pieces of first-class mail between processing centers. Assuming a 40/60 split in traffic by mode and that modal energy intensities are 3,300 Btus per ton-mile for belly freight carried on passenger flights and 2,700 BTUs per ton-mile for intercity trucks, it can be estimated that 1,176 billion BTUs were required to move 53.7 billion pieces.[15] Directly scaling this estimate to match output of the electronic system suggests that 30 billion messages required 657 billion BTUs. On a per-message basis, intercity transport of each message required an average of 21.9 Btus.

When the transfer of 30 billion messages by an electronic system is compared to current conventional transportation practices, only about 571 billion Btus (or 19 Btus per message) could be saved by using electronic message transfer. Such energy savings per year are equivalent to about 98,450 barrels of crude oil. If the energy intensities of both conventional and electronic transfer technologies could be improved by 10 percent over the next decade, the conservation potential would amount to only 88,638 barrels. Actual savings of crude petroleum could be greater than these figures indicate because electronic message-transfer systems use electricity, which can be generated from nonpetroleum sources. By shifting to electronic message transfer, message movement could be accomplished with far less reliance on petroleum.

In terms of the aggregate domestic energy consumption, the potential savings resulting from electronic message transfer are quite small. In 1977, energy use was estimated to total almost 76×10^{15} Btus. Of this aggregate amount, petroleum usage accounted for approximately 37×10^{15} or about 6.7 billion barrels.[16] The energy savings resulting from an electronic system would amount to only 0.5 percent of petroleum consumed in one day and an even smaller proportion of the total energy used in 1977. Yet, an annual savings of 98,450 barrels will not be completely insignificant as petroleum supplies become more scarce.

One might expect much greater energy savings if conventional messages shifted to an end-to-end electronic message-transfer system. However, USPS uses only 350 million gallons of gasoline and diesel fuel per year to move all classes of mail and parcels. This figure, which includes fuel used in all owned, leased, and contracted vehicles, amounts to only 0.001 percent of national petroleum consumption in 1977. Although the implementation of electronic message-transfer technology has the potential to save precious petroleum resources, it cannot offer the resource-saving potential of, for example, transportation or space heating conservation programs. However, it can contribute to general energy conservation efforts. Much greater energy savings are possible as consumers learn to substitute electronic alternatives for a wide range of travel needs. As petroleum supplies become scarcer, the government may well encourage the use and development of electronic message-transfer services.

Vehicle Fleet and Paper Requirements

It is difficult to judge the effect that electronic message transfer will have on postal-owned or leased and contract vehicles. Over the short term, this fleet of 190,000 vehicles will remain relatively stable. Later vehicle requirements may change more radically. For instance, the delivery fleet (about 155,000 vehicles in 1978) may shrink if patrons overwhelmingly adopt electronic collection and delivery options and also if they are willing to accept reduced conventional services (such as three deliveries per week). Intercity contract vehicle needs may also change. If messages migrate from conventional to electronic alternatives, both the size and the number of trucks required may be reduced. Any shift in vehicle fleet requirements will depend on public response to *both* conventional and electronic alternatives.

Electronic message transfer may also have implications for the consumption of paper, the historical medium for recording and storing messages. End-to-end electronic message transfer and office automation systems have been commonly depicted as catalysts for promoting the paperless society. Ideally, recipients could review their correspondence in video form and decide whether to produce a paper copy or simply store it electronically. Considering the amount of paper-based correspondence and transactions flowing among business and government, the potential reduction in paper resources is great. The hybrid electronic message-transfer technology could actually increase use of paper, however, because each time a message is sent from a public terminal two paper copies of the message will be created. Although the technology offers real options for paper conservation, the actual amount will depend on user habits and the system.

Summary

USPS could offer an alternative message-transfer service that may be cheaper, faster, and conserve more resources than conventional services. But along with such developments, USPS will have to alter its mix of capital, labor, and energy resources, change the composition of its labor force, and incorporate expertise, experienced personnel, and electronics into its operations. If the consuming public readily accepts electronic services, USPS may then be able to offer a competitive nationwide system—an option that may some day affect its very viability.

Notes

1. Ronald B. Lee, ''The U.S. Postal Service,'' in *Urban Commodity Flow,* Special Report 102 (Washington, D.C.: Highway Research Board, 1971), pp. 2–40.

2. Morton S. Baratz, *The Economics of the Postal Service* (Washington, D.C.: Public Affairs Press, 1962), p. 1.

3. U.S. Senate, *Evaluation of the Report of the Commission on Postal Service* (Washington, D.C.: U.S. GPO, 1977), p. 74.

4. Douglas K. Adie, *An Evaluation of Postal Service Wage Rates* (Washington, D.C.: American Enterprise Institute for Public Policy Research, 1977), p. 23; Commission on Postal Service, *Report* (Washington, D.C.: U.S. GPO, 1977), vol. 2, p. 649.

5. U.S. Postal Service, *The Necessity for Change* (Washington, D.C.: U.S. GPO, 1976), p. 6.

6. Commission on Postal Service, *Report,* vol. 2, p. 637.

7. U.S. House of Representatives, *Research and Development into Electronic Mail Concepts by the USPS* (Washington, D.C.: U.S. GPO, 1977), p. 3.

8. Charles C. McBride, ''Post Office Mail Processing Operations,'' in *Analysis of Public Systems* ed. Alvin W. Drake (Cambridge, Mass.: MIT Press, 1974), p. 277.

9. U.S. House of Representatives. *General Oversight and Postal Service Budget* (Washington, D.C.: U.S. GPO, 1977), p. 67.

10. Husain M. Mustafa, *The Mechanization and Automation of the*

United States Post Office (Washington, D.C.: Center for Technology and Administration, 1964), p. 40.

11. U.S. Department of Justice, *Changing the Private Express Laws* (Washington, D.C.: U.S. GPO, 1977), p. 29.

12. A.H. Meyburg and A.M. Lee, *An Exploratory Analysis and Assessment of Electronic Message Transfer* (NSF Grant PRA 78–21171, 1982), p. G–27.

13. Mustafa, *Mechanization and Automation,* p. 40.

14. RCA Government Communications Systems, *Electronic Message Service—System Definition and Evaluation* (Washington, D.C.: NTIS, 1978), executive summary, p. 8.

15. G. Kulp and others, *Transportation Energy Conservation Data Book,* 4th ed. (Washington, D.C.: NTIS, 1980), p. 1–28.

16. D.B. Shonka (ed.), *Transportation Energy Conservation Data Book,* 3rd ed. (Washington, D.C.: NTIS, 1979). p. 2–8.

6 Labor Implications

New developments in message transfer will have repercussions for the sources and users of information as well as its processors and movers. Innovative technology could affect the size and composition of this very large aggregate work force, workplace conditions, job design, and the current balance of managerial control and worker power. In this chapter the organizational motivations for adopting this technology and its labor-force implications are considered, particularly the effects on individuals in involved postal and office activities.

At USPS, automation and mechanization may be the key to achieving a variety of management objectives and congressional mandates that include increasing the efficiency of the production process, stabilizing the unit cost of service, and controlling budget growth. Since the reorganization, USPS has been under considerable pressure to provide dependable, low-cost services at wage rates comparable with those paid in the private sector. As salaries have risen in response to the inflation of the last decade, USPS has tried to control the total costs of this labor-intensive operation by increasing output per worker and by streamlining the workforce in an attempt to avoid a projected letter rate of 28 cents in the next 5 years.

Over the last 20 years, through an extensive program to mechanize sorting operations and motorize delivery routes, management has been able to improve productivity by almost 20 percent, at a plant and equipment cost of $100 million for every percentage point of productivity gain.[1] These actions have helped to limit the size of the workforce in spite of growing mail volume. Projections indicate that electronic services may result in much greater productivity gains.

Beyond the projected gains in labor productivity (which are treated in more detail later in this chapter), postal management suggests several other benefits of electronic message transfer. USPS hopes to modernize conventional operations and to offer speed and cost advantages comparable with services of competitors. Personnel administrators hope to alter workplace conditions such that processing and handling evolves from a factorylike production process toward a more white-collar–based activity (perhaps improving worker self-image.) Finally, postal officials may gain more management flexibility with a capital-intensive operation. Current union agreements restrict the hiring and firing of all full-time, part-time, substitutes, or temporary workers during periods of shifting workload. The introduction of electronic message transfer might allow management to reduce the practice of new hires to process seasonal peak loads.

The increase of information activities (that is, correspondence, document production, and transfer) in the business sector has been phenomenal. This trend has been attributed to corporate needs to gather decision-making information to encourage and support efficient production. These activities may allow the market to operate more smoothly. Thus corporate pressures exist for tighter administration with more rapid communication at lower costs. Information handling typically consumes between 5 and 30 percent of an organization's total expenses, occupying more than 20 percent of the domestic labor force.[2] Electronic transfer and office automation technology are being designed to meet these needs.

Over the past decade, industrial productivity has doubled while the productivity of office workers has increased only 4 percent.[3] With costs of technology dropping at a rate of 17 percent annually while labor costs rise 5 percent per year, the terminal in the office has emerged as a promising candidate to reduce the unit cost of information and correspondence, increase communication speed, and reduce paper waste and office requirements for storage. A Booz-Allen study predicts that if office-based white-collar activities reach $1,600 billion over the next decade, new office technology could save over $300 billion.[4]

There may be compelling economic reasons to introduce electronic message-transfer technology into postal and office workplaces. But for reasons of social efficiency, implementation decisions should not be based solely on perceived cost savings or productivity gains. From a wider social perspective, additional benefits are possible if new technology does not diminish worker satisfaction, workplace conditions do not deteriorate, and the effects of job dislocation are mitigated. The following analysis examines the potential effects on labor resulting from electronic message-transfer developments.

The Postal Workforce

Since the reorganization of the Post Office Department and the formation of USPS, postal labor has been a workforce in transition. The introduction of mechanized sorting and motorized delivery routes over the last decade has been accompanied by reductions in force despite an increase in the number of pieces of mail processed and carried. Between 1974 and 1980 the work force fell by 43,000 to 667,000 while mail volume increased 16 billion to about 106 billion pieces. In spite of these reductions, the composition of the employees has remained the same. About 80 percent of the entire staff is made up of full-time employees. In accordance with union agreements, which limit part-time hiring levels, the remaining 20 percent is composed of 90-day temporary substitutes or permanent part-time employees.

Clerks and mailhandlers are the largest group among the full-time employees, with 229,000 workers representing over 40 percent of the regular workforce. Distribution clerks separate incoming and outgoing mail either by hand or machine at postal facilities under close supervision. Their heaviest workloads occur after 5 p.m. when sorting outgoing mail and between midnight and 8 a.m. when sorting incoming mail. These skilled workers, who must meet the strictest job requirements, must demonstrate *scheme* knowledge by passing a state or city examination. A scheme is a large number of destination and distribution points in a bounded geographic area. A grade of 95 percent is required to pass. Other clerks working at customer windows must deal with the public directly, selling stamps or providing other services during common business hours (usually from 8 a.m. to 6 p.m.). All clerks must be familiar with postal regulations and procedures and maintain good public relations with customers. Mail handlers do not deal with the public. They load, unload and move parcels, perform some sorting, and operate some mail-processing machines.

City and rural carriers total about 193,000 workers or 36 percent of the full-time postal labor force. Most of their work is performed semi-autonomously and outdoors during the daylight hours. City letter carriers sort the mail for their routes and deliver it. Rural carriers sort, deliver, and perform duties similar to those of window clerks. Star route carriers deliver mail to remote areas with highly dispersed populations under independent contract with the postal service.

Together, clerks, mail handlers, and carriers make up about 80 percent of the full-time workforce. The remaining 24 percent, apart from supervisory, technical, and administrative staff, is composed of maintenance employees, drivers, and special-delivery messengers. Maintenance employees perform a variety of tasks including elevator operation, equipment repair, and janitorial duties. Drivers pick up and transport mail by truck.

In terms of skill level and training, the clerks are the most highly trained postal labor group. USPS estimates that it takes 6 weeks for clerks to reach minimum standards of productivity. Distribution clerks are not paid to prepare for scheme tests. After passing scheme examinations they require about 2 years to reach peak efficiency. The specialized knowledge acquired by proficient distribution clerks has traditionally enhanced the power of these workers.

Carriers and mail handlers require much less training and skill than distribution clerks. Handlers must demonstrate an ability to lift and move bulky, heavy mail sacks. They typically need only 1 week to rech minimum productivity standards. Carriers require 3 weeks to learn distribution routes. In general USPS estimates that new hires reach a satisfactory performance level after 6 months of employment.

Wage schedules at USPS are set on the basis of four separate pay scales:

The Postal Service Schedule (PS), which sets wages for employees covered by collective bargaining agreements

The Executive and Administrative Schedules (EAS), covering officers, executives, professionals, and administrators

The Non-City Delivery Schedule (NCD), for postmasters in rural areas

The Postal Management Schedule (PMS), for the remaining employees, such as supervisors, technicians, and clerical staff.

These pay scales have been formulated to meet the Postal Reorganization Act mandate that postal workforce wages maintain comparability with other federal service and private-sector salary levels. As additional compensation, USPS provides free life insurance and 75 percent of health insurance payments; insurance, retirement, and other fringe benefits make up 32 percent of the total payroll.[5]

The PS schedule has the greatest effect on Postal Service costs because it covers more than 80 percent of all employees. Table 6–1 compares postal and federal salaries for the 36 postal craft occupations. While clerks and carriers begin at grade 5, some workers reach a grade 6 rating.

Job manuals and letters of instruction precisely describe each job and how it should be performed. Deviations from established rules and procedures by workers can result in suspension, discharge, furlough without pay, demotion, or pay reduction. However, much of the workforce (see table 6–2) is protected by collective-bargaining agreements between the unions and postal management. Additional protection is provided by a grievance procedure, which requires a minimum of 140 days if all levels of appeal are made.

Compensation for the postal workforce has improved greatly since the reorganization, prompting suggestions that these workers may be overpaid. Salary agreements made over the last decade have reduced the time it takes to reach the top of the pay scale from 21 to 8 years. And the quit rate among all workers has fallen to 4 percent of annual hires (1.8 percent for career regulars) as employees have come to value their postal jobs.[6]

Table 6–1
Postal and Federal Salary Comparisons in December 1977

Position	*Grade Level*		*Minimum Salary*		*Difference Over Federal*	*Maximum Salary*		*Difference Over Federal*
	Postal	*Federal*	*Postal*	*Federal*		*Postal*	*Federal*	
Cleaner	PS–1	WG–1	11,710	8,923	2,787	13,613	10,421	3,192
Custodian	PS–2	WG–1	12,129	8,923	3,206	14,186	10,421	3,765
Elevator operator	PS–3	WG–1	12,582	8,923	3,659	14,804	10,421	4,383
Laborer custodian	PS–3	WG–2	12,582	9,464	3,110	14,804	11,045	3,759
Material handling equipment operator	PS–4	WG–6	13,072	11,690	1,382	15,470	13,624	1,846
Warehouseman	PS–4	WG–4	13,072	10,566	2,506	15,470	12,314	3,156
Helper, maintenance trade	PS–4	WG–5	13,072	11,128	1,944	15,470	12,979	2,491
Mail-handler	PS–4	WG–4	13,072	10,566	2,506	15,470	12,314	3,156
	PS–4	WG–5	13,072	11,128	1,944	15,470	12,979	2,491
Tools and parts clerk	PS–5	WG–4	13,604	10,566	3,038	16,189	12,314	3,875
General mechanic	PS–5	WG–8	13,604	12,792	812	16,189	14,934	1,255
Auto mechanic, Jr.	PS–5	WG–8	13,604	12,792	812	16,189	14,934	1,255
Motor vehicle operator	PS–5	WG–7	13,604	12,230	1,374	16,189	14,290	1,899
Carpenter	PS–6	WG–9	14,175	13,354	821	16,980	15,579	1,401
Engineman	PS–6	WG–10	14,175	13,915	260	16,980	16,245	735
Maintenance electrician	PS–6	WG–10	14,175	13,915	260	16,980	16,245	735
Auto mechanic	PS–6	WG–10	14,175	13,915	260	16,980	16,245	735
Tractor trailer operator	PS–6	WG–8	14,175	12,792	1,303	16,980	14,934	2,046
Storekeeper, Auto parts	PS–6	WG–4	14,175	10,566	3,609	17,019	12,314	5,505
Body and fender repairman	PS–7	WG–10	14,794	13,915	879	17,019	16,245	1,574
Machinist	PS–7	WG–10	14,794	13,915	879	17,019	16,245	1,574
Clerk typist	PS–4	GS–3	13,072	7,930	5,142	15,470	10,306	5,164
Telephone operator	PS–4	GS–3	13,072	7,930	5,142	15,470	10,306	5,164
Card punch operator	PS–4	GS–3	13,072	7,930	5,142	15,470	10,306	5,164
Time and attendance clerk	PS–5	GS–2	13,604	7,035	6,569	16,189	9,150	7,039
Clerk-stenographer	PS–5	GS–4	13,604	8,902	4,702	16,189	11,575	4,614
Personnel clerk	PS–5	GS–4	13,604	8,902	4,702	16,189	11,575	4,614
Window clerk	PS–5	GS–4	13,604	8,902	4,702	16,189	11,575	4,614
Distribution clerk	PS–5	GS–4	13,604	8,902	4,702	16,189	11,575	4,614

City letter carrier	PS–5	GS–4	13,604	8,902	4,702	16,189	11,575	4,614
Claims and inquiry clerk	PS–5	GS–7	13,604	12,336	1,268	16,189	16,035	154
Accounting clerk, Intr.	PS–5	GS–7	13,604	12,336	1,268	16,189	16,035	154
Vehicle dispatcher	PS–6,7	GS–7,8	14,175	12,336	1,839	17,819	17,757	62
Electronic technician I	PS–8	GS–5	15,463	9,959	5,504	18,443	12,947	5,496
Electronic technician II	PS–9	GS–7	16,187	12,336	3,851	19,085	16,035	3,050
Electronic technician III	PS–10	GS–9	16,949	15,090	1,859	20,081	19,617	464

Source: U.S. General Accounting Office, *Comparative Growth in Compensation for Postal and Other Federal Employees since 1970* (February 1, 1979), p. 46.

[a]Salary for Federal WG grades is taken from Federal Wage System National Average Schedule that represents a simple average for 135 area Wage Schedules.

Table 6–2
Postal Union Strength

	Postal Employees Having Union Dues Deducted July 1967	*Estimated Postal Union Membership*		
		1966—1967	*1968*	*1975*
American Postal Workers Union, AFL-CIO				260,000
United Federation of Postal Clerks	120,669	162,500	143,000[a]	
National Association of Letter Carriers, AFL-CIO	153,054	201,000	190,000	232,255
National Association of Special Delivery Messengers	2,207	2,400	2,500[b]	
National Rural Letter Carriers Association	2,851	43,000	40,000	50,205
National Post Office Mail Handlers, Watchmen, Messengers and Group Leaders, Division of Laborers' International Union of North America, AFL-CIO	16,340	41,180	35,000	47,000
National Federation of Government Employees	29			1,000
National Alliance of Postal and Federal Employees	23,452	32,000	32,000	45,000
National Federation of P.O. Motor Vehicle Employees	5,972	8,500	8,000[a]	
National Association of P.O. and General Services Maintenance Employees	4,223	21,500	21,500[a]	
National Postal Union	56,369	65,000	70,000[d]	
National Association of Postal Supervisors	26,977	32,000	32,000	34,690
National League of Postmasters	8,079	18,000	18,000	12,000
National Association of Postmasters of the U.S.	9,729	29,000	29,000	28,203

Source: Douglas, Adie, *An Evaluation of Postal Service Wage Rates* (Washington, D.C.: American Enterprise Institute for Public Policy Research, 1977), p. 34.

Note: Between 70 and 90 percent of all employees are unionized. Unionization is much higher among city carriers (98 percent) and rural carriers (92 percent) than among the other employe groups.

[a]The United Federation of Postal Clerks has become the American Postal Workers Union.

[b]Employees in the National Association of Special Delivery Messengers are represented by the National Association of Letter Carriers.

[c]Employees represented by the National Federation of P.O. Motor Vehicle Employees and the National Association of P.O. and General Services Maintenance Employees are represented by other unions, particularly the American Postal Workers Union.

[d]Many New York postal workers represented by the National Postal Union have joined other unions, particularly the American Postal Workers Union.

The demographic characteristics of the postal workforce are very interesting. The average postal worker is male, white, and middle-aged with a high-school diploma. But the detailed statistics are more revealing. Less than 20 percent of the total workforce is female, although 32 percent of postal clerks and 10 percent carriers are women. The median age of employees is very high; almost one-half of the current regular workers are projected to be eligible for retirement during the next decade. About 260,000 workers are over the age of 50 and 150,000 are between 40 and 49.[7]

Data from the late 1960s indicate that while minorities make up only 20 percent of the workforce, postal facilities in the ten largest cities have about 50 percent minority employment. However, minorities mainly hold the more menial positions. The Postal Service has always hired employees with little formal education. In a sample of workers hired during the 1940s, only 1.4 percent were college graduates. A 1966 sample of new hires contained only 3.7 percent college graduates. In recent years, USPS has indicated a desire to attract more college-educated personnel.

Electronic Message Transfer in the Postal Workforce

The size of the future postal work force will be affected by message volume (both conventional and electronic), service standards, and the achievable productivity gains. The Government Accounting Office suggests that labor requirements for the year 2000 could be reduced by 88,000 workyears, if the current level of productivity gains is sustained.[8] Further employee reductions could be achieved if service standards are altered (for example, from six to five deliveries per week) or if conventional messages do not continue to increase at their current rate. Changes in the size and composition of USPS workforce will also be affected by electronic message-transfer developments.

Several studies of postal labor-force requirements for the year 2000 forecast the effect of additional electronic message-transfer developments. Assuming that present service standards are maintained and a 3 percent increase in volume continues, these studies suggest that between 60,000 and 125,000 additional workyears could be eliminated due to electronic developments.[9] The largest reductions would occur in mail-handler and clerk positions. The actual number of eliminated positions will depend on the popularity of electronic services and whether USPS offers all-electronic, hybrid, or only conventional services.

Over the next decade, USPS hopes to develop a mature hybrid electronic system that will partially automate the handling, sorting, and transfer of up to 25 billion pieces of first- and third-class mail. RCA has made recent estimates of the skill level and number of workers necessary to operate such a system. Comparable estimates of conventional services requirements are developed and compared to the RCA projections to assess the implications of the hybrid system on the USPS workforce.

RCA, the current prime contractor for the hybrid-system design, feels that present USPS personnel could be retrained to handle routine operating and maintenance functions. Table 6–3 illustrates one perception of the PS salary grades necessary to staff a typical work station. Mail handlers and clerks could be used to load and unload machines while electronics technicians and equipment mechanics have been assigned to maintenance duties. RCA also suggests that workers with PS–7, PS–8, and PS–9 grades could monitor system operations and perform supervisory duties. Table 6–4 summarizes the staffing requirement of a typical station per shift by salary grade (without supervisory or relief personnel).

By extrapolating this analysis over 87 nonuniformly sized stations, RCA has projected that 4,387 workers drawn from PS grades 4 through 9 could run and maintain a system transferring 25 billion messages annually (see table 6–5). Again, these staffing figures do not include supervisory, overhead, or relief personnel.[10]

To reach a comparable estimate of staffing requirements to process and transfer 25 billion pieces conventionally, several assumptions are necessary. In 1980, USPS employed 229,000 clerks and mailhandlers (in PS grades 4 to 6) to process 106 billion pieces. Assuming that such productivity is maintained through the next decade and because messages can be processed more quickly than other mail items, conservative projections indicate that between 45,000 and 56,000 clerks and mailhandlers could handle an annual load of 25 billion messages. Assuming 5 percent larger workforce for maintenance (such as general mechanics in grade PS–5 and maintenance electricians in grade PS–6), conventional staffing requirements reach between 48,000 and 59,000 workers. Supervisory and overhead personnel would, of course, add more high-level personnel. Because of difficulties in projecting future needs for this labor group in both the conventional and hybrid cases, only craft employees necessary to operate and maintain both systems are included in this comparison.

In general, conventional transfer activities are not as labor-intensive as processing activities. Intercity hauls of first-class mail are mostly carried on air passenger flights or sometimes by truck over shorter distances. Third-class bulk mail travels mostly by rail. Contractors, not postal employees, handle long-haul transportation. Mail is moved among intraurban facilities and transportation terminals by 11,000 postal employees (55 percent PS–5 drivers and 45 percent vehicle service PS grades 5 and 6). But these jobs are somewhat volume-independent, because freight runs are scheduled according to delivery performance standards

Table 6–3
Staff Work Load—Typical EMSS Station

	Fraction of Workers' Time		*Salary Grade*	
	Operations	*Maintenance*	*Operations*	*Maintenance*
Magnetic tape drives	0.03	0.03	5	8
Magnetic card readers	0.20	0.16	4	8
OCR FAX	2.00	1.00	6	8
Forms scanners	1.00	0.12	5	8
EMBs (public terminals)	8.2[a]	8.2	4	9
Printers	10.8	7.38	4	8
Computers	–	0.45	–	9
Letter mergers	1.75	0.42	5	7
Front-end controllers	–	0.08	–	8
Random access storage	–	0.28	–	8
High-speed storage	–	0.77	–	8
Satellite ground terminal	–	0.24	–	8

Source: RCA Government Communications Systems, *Electronic Message Serivce*—(final report), p. 3–62.
[a]Collecting tapes from EMBs is costed as an operator function.

Table 6–4
Staffing Requirement—Typical EMSS Station

Operator		*Maintenance*	
Salary Grade	*Number*	*Salary Grade*	*Number*
4	20	7	1
5	3	8	11
6	2	9	9
Total	25		21

Source: RCA Government Communications Systems, *Electronic Message Service* (final report), p. 3–63.

Table 6–5
EMT Personnel Required for Processing 25 Billion Messages per Year (Hybrid System)

Grade	*Number*
4	1,943
5	460
6	318
7	87
8	831
9	748
Total	4,387

Source: RCA Government Communications Systems, *Electronic Message Service,* (final report), p. 4–10.

Table 6–6
Staffing Requirements for Processing 25 Billion Messages per Year (Conventional System)

Grade	*Range*
4	12,000–15,000
5	25,580–30,660
6	13,520–16,540
Total	51,100–62,200

and vehicles leave at specific times, regardless of load factors. Yet, to include a reasonable contribution of all postal labor to transfer activities, 100 to 200 workers are assumed to be responsible for 25 billion messages.

Tables 6–5 and 6–6 show possible differences in staffing requirements (without supervisory or overhead personnel) for an electronic hybrid system and a conventional postal system, each processing and transferring 25 billion messages annually. It is possible to reduce staffing in these activities tenfold or about 50,000 workers by substituting the hybrid technology for conventional equipment and labor. Under the hybrid system about 45 percent of the new positions will be scaled at PS grade 4 while the remaining jobs will be distributed among PS grades 5 through

9. By comparison, the conventional system has jobs clustered in PS grades 4, 5, and 6, with most jobs at the grade 5 level.

The movement to a hybrid system, then, involves a large-scale reduction in force and a replacement of workers mostly by lower-grade workers and also by more skilled workers (fewer in absolute terms). Beyond the next decade, if the hybrid service were to capture a greater portion of the first- and third-class message markets, similar reductions in employment might be expected. In achieving the projected diversion of mail, if customers choose to use end-to-end systems over a USPS hybrid service or if USPS is prevented from offering a hybrid service, then the reductions in postal positions could be even greater. In all cases, the actual number of positions eliminated will probably be negotiated by the parties involved.

In the event that USPS develops an all-electronic system or if the hybrid system evolves in that direction, employment impacts would be difficult to estimate. For instance, the carrier force, currently composed of 193,000 workers, would certainly be affected. However, changes in the size of the delivery force will depend more on the future popularity of conventional services, number of delivery days per week, and number of households or businesses serviced rather than changes in volume.

Many observers find comfort in the possibility that potential reductions in the postal workforce can be achieved through attrition, implying that layoffs will not be necessary. RCA, for example, notes that "attrition via retirements and reduced recruiting will achieve this reduction with little or no impact on individual carriers."[11] Indeed, demographic statistics reveal that more than half of the current full-time employees could retire during the next decade. But demographic characteristics also give some indication of the role that postal employment has played in the lives of upwardly mobile but disadvantaged groups in our society. Postal jobs have traditionally served as a means for veterans, poorly educated workers, and minorities to gain reasonable and secure employment when other opportunities have been absent. Although the projected workforce reductions may not displace current employees, future generations of disadvantaged groups may lose these opportunities for economic advancement.

Electronic systems have other potential effects on the postal workforce beyond the issue of workforce reduction. The hybrid technology has the potential to free distribution clerks and mailhandlers from the routine and arduous tasks of memorizing schemes, sorting, and moving 70-pound mail sacks with great frequency. Yet some workers fear such repercussions as the loss of wages or bargaining power if they lose the unique power that scheme knowledge bestows. More radical labor elements theorize that management's willingness to incur such a large fixed capital investment (about $1.8 billion) is motivated by a desire to reduce the possibility of another nationwide postal strike by the most militant labor force in the public sector.

Blauner's characterization of worker alienation suggests that there are implementation questions which will have substantial impact for postal labor.[12] Electronic message-transfer technology offers the potential to make the workplace exhibit more of the conditions that prevail in white-collar work places and fewer factory like conditions. But job tasks and the organization of jobs should reflect attempts to increase worker autonomy, challenge, and responsibility. The electronic message-transfer workplace could provide variety and improve occupational skills and personal growth. Without adequate attention to the potential effects listed here, as well as such ideas as job rotation, job enlargement, and job enrichment, USPS may face problems arising from worker dissatisfaction (such as lower productivity, poor morale, sabotage, or turnover) that mitigate against the expected benefits.

The Office Workplace

The office workplace tends to be a varied setting where knowledge, information, and perceptions are conceived, assimilated, and communicated; where decisions are made and implemented; and where documents and messages associated with these activities are created, processed, and stored. The character of this workplace as well as the size, composition, and skill of the workforce has changed dramatically over the last decade. Before discussing the effects of electronic message transfer and its supporting technology on office workers, the characteristic features of this workforce and workplace are briefly described.

Office workers make up about 48 percent of the total national workforce. Managers, professionals, and office-based white-collar personnel can be categorized in three groups: principals, secretarial employees, and clerical workers.

Principals, consisting of upper management, middle management, and professionals, comprise over 50 percent of the total office workforce and account for 70 percent of office labor costs. These highly paid workers make business decisions, attend meetings, delegate responsibilities, conduct administrative tasks, and travel. Table 6–7 summarizes the activities of these workers in a typical office. The communication load (including reading, writing, meeting, traveling, and telephoning) of these highly skilled workers accounts for over half of their work activities; this proportion is even higher for upper-level management. This group is primarily composed of college-educated, white males.

Table 6–7
Principal Worker Activity

	(average percent of time)			
Activity	*Level 1 (Upper Managers)*	*Level 2 (Other Managers)*	*Level 3 (Non-managers)*	*Total*
Writing	9.8	17.2	17.8	15.6
Mail handling	6.1	5.0	2.7	4.4
Proofreading	1.8	2.5	2.4	2.3
Searching	3.0	6.4	6.4	5.6
Reading	8.7	7.4	6.3	7.3
Filing	1.1	2.0	2.5	2.0
Retrieving filed information	1.8	3.7	4.3	3.6
Dictating to secretary	4.9	1.7	0.4	1.9
Dictating to machine	1.0	0.9	0.0	0.6
Telephone	13.8	12.3	11.3	12.3
Calculating	2.3	5.8	9.6	6.6
Conferring with secretary	2.9	2.1	1.0	1.8
Scheduled meetings	13.1	6.7	3.8	7.0
Unscheduled meetings	8.5	5.7	3.4	5.4
Planning or scheduling	4.7	5.5	2.9	4.3
Traveling outside HQ	13.1	6.6	2.2	6.4
Copying	0.1	0.6	1.4	0.9
Using Equipment	0.1	1.3	9.9	4.4
Other	3.1	6.7	11.4	7.7
Total	100.0	100.0	100.0	100.0

Source: G.H. Engel and others, "An Office Communications System," *IBM Systems Journal* 18 (3), 1979, p. 403.

Secretaries and typists, whose wages account for only 6 percent of office labor costs, can be roughly divided into two groups according to activity assignments. Private secretaries who work for only one boss spend only a small fraction of their time typing (typically one-quarter of their work time). Other activities include conferring with the boss, keeping a calendar, taking shorthand or telephone messages, and handling mail. For less-specialized secretaries who work for a number of people, typing duties tend to consume almost one-half of their work time; the remaining time is devoted to other assorted duties. Table 6–8 is an aggregate listing of activities and an average division of job responsibilities. Secretarial workers usually possess typing, file-handling, editing, and telephone reception skills that are less valued than those of principals. Demographic statistics show these workers to be predominantly white females with high-school degrees.

Clerical workers make up the remaining office labor force and tend to be the lowest-paid office workers. Their duties consist of a medley of tasks that are not assigned to secretaries or principals. Table 6–9 is a

Table 6–8
Secretarial Activity

Activity	*Average Percent of Time*
Writing	3.5
Mail handling	8.1
Bulk envelope stuffing	1.4
Collating/sorting	2.6
Proofreading	3.9
Reading	1.7
Typing	37.0
Telephone	10.5
Copying or duplicating	6.2
Conferring with principals	4.3
Taking shorthand	5.5
Filing	4.6
Pulling files	2.8
Keeping calendars	2.6
Pick-up or delivery	2.2
Using equipment	1.3
Other	2.0
Total	100.0

Source: G.H. Engel and others, "An Office Communications System," *IBM Systems Journal* 18 (3), 1979, p. 404.

Table 6–9
Clerical Activity

Activity	*Average Percent of Time*
Filling out forms[a]	8.3
Writing[a]	7.3
Typing[a]	7.8
Collating/sorting[a]	5.2
Checking documents[a]	10.4
Reading[a]	2.9
Filing[b]	5.9
Looking for information[b]	10.2
Telephone	9.2
Copying or duplicating	3.9
Calculating	10.3
Meetings	1.9
Pickup or delivery in HQ	0.8
Scheduling or dispatching	1.2
Using a terminal	6.3
Other	8.4
Total	100.0

Source: G.H. Engel and others, "An Office Communications System," *IBM Systems Journal* 18 (3), 1979, p. 406.

[a]Primary paper-handling activities (41.9 percent)

[b]Secondary paper-handling activities (cumulative total = 58 percent)

sample of clerical activities along with typical allocations of work time. These positions tend to be primarily focused on paper-handling activities. In general, clerical workers are considered to be the least skilled of all office workers. Like secretaries, most clerical workers in offices are white, female, and hold high-school diplomas.

Most analyses of working conditions in the office have noted a trend toward increasingly routine and specialized tasks, automation of most repetitive tasks, and a reorganization of clerical functions into a factorylike process.[13] Changes in both the form and content of office work have been influenced by technological change. Four evolutionary stages of development characterize this trend.[14]

1. *Craft accurate work*—In this stage, office work consisted primarily of the accurate and rapid shorthand of the private secretary, the numerical accuracy of the bookkeeper, and the clear, careful penmanship of the ledger clerk. Office workers considered these tasks to be integrally related to the management of the organization, and acquired skills and knowledge valuable for occupational mobility.
2. *Early mechanization*—Typewriters, dictating machines, adding machines, and other office machines were introduced in this stage. The craft aspect of work was reduced, and the distinction between managerial and clerical staff became more pronounced. With the formation of typing pools, factory conditions made their first appearance.
3. *Punched-card data processing*—Early data-processing technology created some new tasks and eliminated others. Such new semiskilled tasks as punching, verifying, and business-machine operation were introduced to the office. The data-processing equipment replaced some skills, such as the use of adding machines in offices that handled large amounts of records and information.
4. *Electronic data processing*—Computer programs, systems analysis, operators, and maintenance personnel were introduced to supervise and operate data-processing functions. New semiskilled tasks were created to prepare data for the computer. Introduction of this technology tended to increase the centralization of data-processing functions.

In the contemporary office of today, job responsibilities still constitute a mixture of these activities.

Because the Bureau of Labor Statistics collects data only for national organizations, there are no comprehensive data on union membership by occupational group. Other sources reveal highly fragmented efforts to unionize office workers. Principals have never sought unionization. Only

2 percent of the offices with secretaries and clerical workers are completely unionized, while 6 percent are partially unionized.[15] Nonbargaining groups, such as the National Secretaries Association, as well as local, state, national, and international unions, such as the Service Employees International Union, have attempted to increase membership and visibility. And large unions, such as the Teamsters, United Autoworkers, or the American Federation of State, County and Municipal Employees, are increasing their efforts in the office workplace. However, organized labor has not enhanced the power of office workers.

There are very real differences among the three office-worker groups. Principals are highly skilled, highly educated decision makers who are well compensated and have much job mobility. For the two other groups, office employment typically represents a way for young, less-educated workers with few or no skills to earn a living. In the current office workplace, upwardly mobile workers can upgrade their skill levels (for example, from clerk to typist, to secretary, to private secretary, to administrative assistant). But job security for secretaries and clerical workers is not as high as for postal workers. In the future, increased unionization of office workers may change these employment conditions.

Electronic Message Transfer in the Office Workplace

The assertion that introducing innovative information-handling and message-transfer technology into the office workplace will affect all categories of office workers is largely unquestioned. There is, however, much disagreement concerning the magnitude and nature of these effects. Some suggest that significant productivity gains can be achieved by developing word-processing centers that are staffed by clerk-typists. Others warn that the major impact of new office technology will be achieved by having executives augment their decision-making abilities with computer power delivered by message terminals. The combination of new message-transfer, text production and handling, teleprocessing, and information-retrieval capabilities in the office workplace may act as a powerful agent of social change for this community of workers.

Original development of the word-processing concept began at IBM of West Germany in the early 1960s. Since that time, various German and U.S. studies have attempted to estimate the productivity gains which could be achieved by automating typing, editing, handling and transfer activities associated with office work.[16] The Siemens company has estimated that by 1990 about 40 percent of present office work could be transferred to word-processing, causing trade unions to suggest that two million jobs could be lost from Germany's projected labor pool of five million secretaries, typists, and clerical workers. Another study by a German federal commission suggested that a savings of 35 percent were possible while a U.S. study estimated that six word-processing operators could replace fifteen workers. Although there is a lack of agreement over achievable productivity gains, nearly all studies assert that positive gains are likely.

Several examples illustrate the potential impact of increased labor productivity on national office employment levels. Considering the divergent productivity claims, a conservative assumption that three word-processing operators could do the work of four full-time typists might be made. If such productivity gains were applied to the 1977 national total of 1 million typists, 250,000 full-time typing positions might be eliminated if output remained the same. In periods of expanding output, new hires could be avoided, thereby avoiding the problems of layoffs of already employed workers. In addition, because word-processing or electronically transferred text and messages are amenable to automated filing, storage and retrieval, productivity gains among file clerks can be anticipated. If similar productivity gains were achieved for file clerks (that is, three for four) then 274,000 positions existing in 1977 could be handled by 205,000 workers, representing a reduction of 69,000 full-time clerks at constant ouput.

Estimating employment level impacts of electronic message transfer and word-processing technology on secretarial personnel is more difficult because of the variety of their tasks. Exemplary calculations can again illustrate the nature of the effect. Secretaries typically spend one-half of their time typing and handling files and the remaining half on other responsibilities (see table 6–8). Some management consultants concerned with implementing word-processing have recommended that secretarial jobs be restructured so that some workers would handle only files and typing while others would become administrative assistants. In that event, assuming again that three word processing operators could handle the job of four clerk-typists, three word-processing operators plus four administrative assistants could handle the work of eight secretaries at constant output. Applying this reasoning to the 3.4 million secretarial jobs existing in 1977, one-half of these positions could be upgraded to low-level administrative jobs while the remaining 1.7 million positions could be downgraded and reduced to 1.275 million clerk-typist positions.

If electronic message transfer and word-processing were implemented in all offices in 1977 successfully and if three word-processing operators could do the work of four clerk-typists, as many as 744,000 positions might be eliminated while three million jobs could be upgraded or downgraded with output unchanged. It is also possible that hardware costs will drop to a point where equipment will be widely available to most office workers. In that case, skill levels and personnel levels may remain the same while the allocation of secretarial work tasks may change. The actual changes in job structure and office positions will depend on the implementation plan, each office setting, and hardware costs.

Because technical personnel could be required to maintain word processing equipment (although leased equipment is often serviced by leasing companies), a few of these eliminated positions could be saved and upgraded to skilled, technical, or engineering jobs. Also, mail-room positions and intracompany messenger labor in larger corporations could be eliminated if more word-processors are equipped with communications or message-transfer capabilities. Of course, any productivity gain could result in no job losses for the current workforce by avoiding new hires as output rises.

Up to now, word-processing has not become a significant factor in the office workplace for a variety of reasons. One knowledgeable observer has commented that word processors "costing $10,000 to $20,000 could hardly be cost-justified as simple replacements for $850 typewriters, regardless of their promise."[17] Others feel that the lack of success of past implementation efforts can be attributed to poorly designed equipment, the alienation resulting from factorylike installations of machinery, lack of attention by management after installation, an unwillingness of executives to give up private secretaries, inadequate restructuring of office activities to accommodate equipment, and a variety of other problems. Based on these experiences, it is difficult to be sure that word-processing equipment will be widely used in the future and even harder to make a detailed analysis of the resulting effects of this use.

The future of word processing will look much brighter if equipment costs continue to decrease as expected and if lessons are learned from past implementation experiences. During the next decade, word processing could result in a 60 percent savings over conventional typing when costs of corrections and redrafts of documents or correspondence are included in calculations (now estimated to cost $4.50 per page).[18] If this productivity is combined with potential speed and cost advantages of electronic message-transfer and data-processing capabilities, the attractiveness of such equipment could increase even further.

If word-processing activities facilitate office productivity gains and if reductions in force are necessary, management will probably attempt to make adjustments through attrition or reassignment to other duties. In several examples of office automation, Williams and Lodahl found that layoffs of office staff were rarely necessary when innovative firms adopted centralized word-processing. Workers tended to fill redefined administrative and clerical supervisory positions or they quit. Most of the lower-level positions were filled with new hires.[19]

For the workers who fill newly redesigned jobs, word processing creates a situation of mixed opportunity. Administrative assistants may find their job responsibilities to be more varied, more challenging, and less routine than secretarial postions. Although word-processor operators may have less repetitive typing and redrafting duties, their prospects for promotion or skill acquisition may be reduced. In addition, they may face constantly heavy workloads with uncontrollable input queues, little work variety, and decreased interactions with other workers. If personnel management trends continue, workers who remain with firms after office automation is introduced face either a future as paraprofessionals or they may have to accept positions with more limited prospects. Alternatively, equipment costs might eventually fall to a point where machine idle time is not particularly costly and skill levels will not change radically. The possible loss of job opportunities together with changing work conditions could make the advent of word processing catalyze office unionization, particularly if workers feel the need to enhance their abilities to deal effectively with management.

Office workplace conditions will be altered by the introduction of word processing. The problems associated with implementing centralized, isolated word-processing centers are well documented in the literature.[20] Employees may find dissatisfaction, for example, from long-term occupational exposures to video units causing eye strain or headaches or from monitoring operators' productivity levels. The success of word-processing activities, and integrated text management and communications systems, will depend on satisfactory reorganization of jobs and office procedures and on the resolution of more individual employee problems resulting from office automation. By increasing worker satisfaction along with labor production, management will avoid problems of alienation, absenteeism, low-quality output, and unnecessary recruiting and training costs associated with high turnover.

Innovative office technology might have significant implications for management, professionals, and executives. Principals may be encouraged to use personal desk terminals for reasons of speed, confidentiality, and efficiency. It has been suggested that high level decision makers and administrators with direct access to database information, communication channels, and computing power might enhance their skills by being able to search for, manipulate, and structure information, thereby increasing their productivity levels. As a result, the most effective productivity gains from new office technology may come from saving time and effort for this highly paid group.[21] However, executives might resist such efforts because they lack keyboard skills or because they are unwilling to give up personal secretaries on whom they depend.

Policy Implications for Labor

To consider possible policy initiatives that address the effects of electronic message transfer and word processing on labor in the postal and office workplaces, one needs to understand various objectives of the involved actors (including workers, management, and government) in the context of larger social goals. Although most actors agree that increased labor productivity is a desirable goal, some argue that a narrow focus on economic criteria (such as output per labor unit) may be suboptimal when considering an overall objective of increasing social efficiency. If productivity increases are accompanied by reduced opportunities for new workers entering the labor market, by layoffs and by alienating work conditions, society at large may incur costs above the benefits derived from increased labor productivity. For instance, health costs associated with job displacement and depressing work conditions plus social costs resulting from mid-career job stagnation of blue-collar and white-collar workers, as well as costs of crime attributed to unemployment, may result in net disadvantages to society. As a result, programs in both the public and private sector may be necessary to counteract the potential negative consequences of increasing labor productivity. Various alternatives are available.

Job Mobility and Retraining

Historically, society has relied on the mobility of labor to adjust to shifting employment conditions in a particular industry. Workers moved to different locations within the same industry or to different industries that could use their skills. Ambitious employees expected to advance as companies provided retraining primarily for low-level processing and supervisory positions. Major retraining costs for technical or managerial positions were traditionally borne by the individual. And these costs for the mid-career worker, which included foregone income and sacrifices by families, tended to be quite high. As a result, significant professional advancement has commonly been an intergenerational process.

Collective Bargaining

Collective bargaining, a radical social concept 100 years ago, is now well established and supported by legal statutes. It has been used by organized labor to define and protect the working conditions of much of the labor force. The success of collective bargaining is based upon a union's ability to organize and control workers along with the ability of management and organized labor to negotiate binding agreements.

Job Enrichment, Enlargement, or Redesign

Various experiments have attempted to restructure the content of jobs and worker responsibilities. These activities incorporate strategies to increase the autonomy of individual workers and work groups, to provide for job mobility and reward the acquisition of new skills, to democratize decision making in the firm, and to construct a supportive, stimulating work environment.[22] The goal of such efforts is to increase the job satisfaction of the individual worker and to reduce worker-management conflicts.

The role of unions, management, and government in addressing the problems of automation has been varied. With regard to mobility and retraining, government has been advised to concentrate upon expanding general educational oportunities and promoting national economic growth to create more jobs.[23] In a few cases, government has enacted legislation to protect groups adversely affected by other federal actions (such as transportation and communication workers). Most of these efforts provided wage maintenance after job loss for periods ranging from one year to retirement age for some displaced workers.[24] Although some officials have endorsed job redesign experiments, government has largely provided only the most general assistance to workers coping with industrial automation (such as employment offices and job programs).

Industry has not been totally unresponsive to the plight of workers facing automation-induced changes. In offices, new strategies of job redesign have been aimed at resolving worker dissatisfaction.[25] At USPS, job opportunities have increased for workers since the reorganization. But, in general, management has not undertaken large-scale retraining programs without the encouragement of organized labor.

Unions have been particularly effective in the postal sector and almost nonexistent in the office workplace. Collective bargaining by postal workers has led to protection of regular employees from loss of income or jobs due to automation. However, their agreements do not protect more senior employees from mid-career stagnation nor are problems of job redesign addressed.

Various actions could be taken to mitigate the undesirable effects of electronic message transfer and word processing in the postal and office workplaces. Because of the public nature of USPS, government might feel compelled to retrain or shift displaced employees into jobs that are not downgraded. In addition, government might expand programs that increase entry-level job opportunities for the disadvantaged. Labor unions and management might be actively encouraged to develop job redesign implementation strategies and job-guarantee programs that minimize employee dissatisfaction. Finally, unions might work to formalize many of these actions in collective-bargaining agreements.

In the office workplace, many expect office-automation technologies such as electronic message transfer to catalyze union organizing. Such development could precipitate collective-bargaining agreements that could benefit both workers and management. Workers may be able to negotiate the nature of working conditions, job design, and retraining responsibilities of employers while management may gain a means to deal with labor conflicts and enhance productivity through automation. The government might expand Department of Labor programs to match skills of displaced workers with available jobs, increase continuing education and vocational training opportunities for full-time workers, and also begin to address the retraining needs of workers in particular industries that are affected by automation. Finally, the government might support experiments in job redesign, job enlargement, and job rotation through demonstration grants to see whether technological changes such as office automation can be successfully implemented in a way that both management and workers are more satisfied with changing employment conditions.

Each of these actions, however, will have limited effectiveness in resolving the wide range of issues arising from automation. Our society has lacked a public-private cooperative effort to provide education and vocational retraining throughout an individual's career to deal with separation from and reentry into the labor force, without profound income, physical and psychological costs. This idea has been recommended by several public commissions studying automation problems. Without some concerted form of public and private intervention to facilitate adjustments, the negative effects are likely to contribute to growing social problems of unemployment, rising health care costs, and other manifestations of job dissatisfaction in the postal and office workplaces.

Notes

1. U.S. House of Representatives, *Research and Development into Electronic Mail Concepts by the USPS* (Washington, D.C.: U.S. GPO, 1977), p. 4.
2. James Martin. *The Wired Society* (Englewood Cliffs, N.J.: Prentice-Hall, 1978), p. 197.
3. Yankee Group, "Making Office Automation Work," *Telecommunications Policy* (June 1979), p. 152.

4. Booz Allen, "Office Automation in the '80s" (mimeo of seminar presentation), 1980.

5. U.S. General Accounting Office, *Comparative Growth in Compensation for Postal and Other Federal Employees Since 1970,* FPCD–78–43 (Washington, D.C., February 1, 1979), p. 17.

6. Douglas K. Adie, "Has the 1970 Act Been Fair to Workers?" (paper prepared for a Conference on Postal Service Issues, October 1978), pp. 81, 109.

7. U.S. General Accounting Office, *Implications of Electronic Mail for the Postal Service's Work Force,* GGD–81–30 (Washington, D.C., February 6, 1981), p. 37.

8. Ibid., p. 26.

9. Ibid. Also see National Research Council, *Review of Electronic Mail Service Systems Planning for the U.S. Postal Service* (Washington, D.C.: National Academy Press, 1981), p. 28.

10. RCA Government Communications Systems, *Electronic Message Service—System Definition and Evaluation* (Washington, D.C.: NTIS, 1978), Final Report.

11. Ibid., executive summary, p. 54.

12. R. Blauner, *Alienation and Freedom: The Factory Worker and His Industry* (Chicago: University of Chicago Press, 1964).

13. J.T. Dunlop, *Automation and Technical Change* (Englewood Cliffs, N.J.: Prentice-Hall, 1962), p. 61.

14. Jon M. Shepard, *Automation and Alienation: A Study of Office and Factory Workers* (Cambridge, Mass.: MIT Press, 1971), pp. 41–47.

15. Administrative Management Society, *AMS Professional Management Bulletin* 14 (June 1976), p. 6.

16. Post Office Engineering Union, *The Modernization of Telecommunications* (London: College Hill Press, June 1979), p. 95; External Telecommunications Executive, *Preliminary Report on Word Processors and International Communications,* Long-Range Studies Division, British Post Office (September 1978), p. 8.

17. Yankee Group, "Making Office Automation Work," p. 152.

18. James E. George, "Word Processing—Present and Future," *Proceedings of AESOP Conference* (Seattle: May 3–4, 1977), p. 7.

19. Lawrence K. Williams and Thomas M. Lodahl. "Comparing WP and Computers," *Journal of Systems Management* 29:2 (February 1978), pp. 9–11.

20. Ibid. Also see Williams and Lodahl, "An Opportunity for OD: The Office Resolution," *OD Practitioner* 10:4 (December 1978), pp. 9–12.

21. R.C. Harkness, "Office Information Systems," *Telecommunications Policy* (June 1978), pp. 91–105.

22. See U.S. Senate, Committee on Labor and Public Welfare, Subcommittee on Employment, Manpower, and Poverty, *Work in America,* 93rd Congress, 1st Session (February 1973), chapter 4.

23. Derek Bok, *Automation, Productivity and Manpower Problems* (Washington, D.C.: U.S. Department of Labor, 1964).

24. U.S. Department of Labor, *Public Policy and Economic Dislocation of Employees* (Washington, D.C.: Office of Assistant Secretary for Policy Evaluation and Research, October 1978).

25. Williams and Lodahl, "Comparing WP and Computers," pp. 9–11.

Appendix 6A Automation and Work: Several Views

Mechanization and automation have been depicted as either a boon or a bane for the worker and society at large. Optimists portray automation as a means to free workers from repetitive or unrewarding labors, as a vehicle for increasing job opportunities and the well-being of the labor force, and as a way to reduce waste and alleviate age-old problems of material scarcity. Social critics have expressed fears that jobs will be lost, especially by middle-aged workers, that craftsmanship will be stifled, and that management will further antagonize workers. Broady focused studies examining automation, productivity, and employment issues, along with research dealing with individual industries, reveal that there is some truth to all of these positions. Several conventional viewpoints seem to prevail.

Under historical conditions of economic growth, it is difficult to judge the net effect that automation has had on employment levels. Job losses have occurred when automation has been introduced into specific industries (such as machine tool production) and new jobs have been created. Across all industries, from a broader perspective, employment levels seem to be tied more closely to the performance of the national economy than to increasing levels of automation.

The changing composition of the labor force, one of the most dramatic trends of the last 20 years, has caused social commentators to speculate about the emergence of a postindustrial society.[1] Census figures over the last two decades indicate a doubling of professional, technical, and skilled worker employment. Proportionately, this occupational group has grown from 11 to 15 percent of the workforce while blue collar employment shrank from 37 to 35 percent.[2] These aggregate statistics, which seem to imply more highly paid workers with better jobs, may reflect the effect of automation. Alternatively, they may be characteristic of a growing service-delivery sector.

Coincident with the increase of professional and technical workers, there has been a dramatic shift in the educational level and training of the workforce. Since the turn of the century, primary and secondary educational attainment among the aggregate population has been rising steadily. Opportunities for higher education increased after World War Two as federal subsidies for war veterans grew during the 1950s and as state legislatures began to expand their university systems. Social theorists suggest that the development of a more highly educated workforce has facilitated the design, management, and operation of automated production processes and as a result society is better off.

Not everyone agrees with the analysis developed from earlier national studies of automation and labor. Recent studies have argued that a dual labor market exists. Those persons in the upper level are well paid, have advancement opportunities, and are able to move fairly easily when faced with technological progress in an industry. Workers in the lower tier, who tend to be disadvantaged minorities, migrants, women, or youth, often find reemployment difficult when faced with industrial automation.[3] Even if the introduction of automation leads to a net gain in jobs, those displaced individuals may not benefit from the change because others, perhaps better educated and from the upper tier, may be trained for and hired into the new positions. Only on an intergenerational basis are some workers able to benefit from greater opportunities.

With respect to the changing composition of the workforce, critics such as Braverman note that the dramatic shift in the skill level of the workforce may be partially a product of arbitrary statistical categorization. He states that by "making a connection with machinery . . . a criterion of skill, it guaranteed that with the increasing mechanization of industry the category of the 'unskilled' would register a precipitous decline, while that of the 'semi-skilled' would show an equally striking rise."[4] Indeed, the workforce may not be better off—the workplace may be merely more mechanized and automated.

Braverman also notes critically that the "idea that the changing conditions of industrial and office work will require an increasingly 'better-trained,' 'better-educated,' and thus 'upgraded' working population is an almost universally accepted proposition of popular and academic discourse."[5] One problem with this point of view is that it uses average values, which can mask increasing polarization occurring within the industry being observed. For instance, both upgrading of some positions and downgrading of others could be occurring simultaneously. Thus a conclusion could be reached that the average skill level is the same or has been gradually increasing.

In general, those who find fault with the findings of earlier automation studies argue for a more disaggregated analysis concerning the nature of automation and its effect on workers and workplaces. Silverman in a 1966 study of automation effects differentiated three categories.[6] Process A replaces the need for human physical strength with the use of machines. Process B involves the mechanization of processes by linking machines in a production line. Process C is described as using computers to carry out clerical or bookkeeping work. Under this characterization, process A decreases or eliminates the need for heavy manual labor while processes B and C change the labor requirements for accomplishing monotonous and repetitive tasks in manufacturing and office contexts. Although this scheme differentiates types of tasks affected by automation, it does not describe automation effects according to changes in employment level, skill of workers, and production output. To relate these characteristics, a typology of the following automation situations has been created:

Type 1—Replacing many unskilled workers by machinery and fewer skilled workers who watch the equipment and make adjustments, to achieve constant or modest increases in output at lower total labor costs.

Type 2—Replacing skilled workers (such as craftsmen) by unskilled workers plus machinery to lower average wage costs per worker.

Type 3—Using same number of workers and skill levels with machinery to increase the pace of production and increase output. A corollary situation would be to use machinery to increase production level and reduce future new hires.

All three types of change to automation could increase labor productivity and lower unit output costs. But each situation has very different implications for the affected workers in terms of changes in skill level, employment opportunities, and the nature of the work process. It is equally important to examine automation-induced changes in the workplace along with shifts in employment levels.

Blauner, in a classic study of worker alienation, identified four principal types of dissatisfaction.[7] He found that some workers feel powerless because they lacked influence or control over work conditions or the production process. Workers also doubted their meaningfulness to the work process in terms of their own responsibility for and contribution to the final product. Another source of dissatisfaction arose from a sense of social isolation because the work process commonly does not promote informal relationships between workers. Finally, workers often have a depersonalized detachment from the work process and regard it only as a means to gain sustenance.

Automation efforts in the workplace may not have been responsive to the problems of worker alienation. One critic notes that "jobs are still organized with the objective of harnessing workers to the needs of the production process rather than vice versa."[8] Historically, the application of the concept of division of labor has reduced the scope of each individual involved in a production process. Such efforts attempt to increase labor productivity without considering the possibilities of downgrading skills of workers (that is, assuming that the reduced subtasks do not require as much skill to perform as the overall complex task) or of routinization of the work task. Arendt asserts that these work processes have been further transformed by automation into new sorts of labor activities.[9] For example, the introduction of automation machinery among workers in the workplace has, in some cases, further delimited individual job tasks and skill requirements, decreased worker autonomy by establishing work pace according to machine capabilities, increased physical and social isolation of workers, and promoted scheduling work on shifts to use machinery more efficiently.

The more severe critics characterize management's use of automation as a blatant conspiracy against workers. For instance, machines could control the pace of worker productivity (this is a Type 3 situation). Alternatively, machines might serve a policing function by monitoring each individual's output and automatically informing supervisors about workers who do not achieve production quotas. In addition, changes in job skill requirements resulting from automation could reduce the earning potential of workers (Type 2). Furthermore, such changes could break management's dependence on specialized knowledge and decrease training costs of new hires. And finally, reductions in work force (Type 1) might make operations easier to run in situations with labor problems. Such situations are not inconceivable and some have occurred.

Thus, automation is neither an inherently benign or an evil force in the workplace. As a tool it can alleviate the need for human physical toil. As a tool it can be used to reduce human efforts devoted to repetitive and monotonous work tasks. But it can also contribute, whether intentionally or not, to increases in workers' alienation or the span of their managerial control. Thus, implementation strategies are of critical importance.

Notes

1. See, for example Daniel Bell, *The Coming of Post-Industrial Society* (New York: Basic Books, 1973).
2. Eli Ginzburg, "The Professionalization of the U.S. Labor Force," *Scientific American* 240:3 (March 1979), p. 62.
3. M.J. Piore, "Notes for a Theory of Labor Market Stratification," in *Labor Market Segmentation,* ed. R.C. Edwards and others, (Lexington, Mass.: Heath, 1981).
4. Harry Braverman, *Labor and Monopoly Capital* (New York: Monthly Review Press, 1974), p. 429.
5. Ibid., p. 424.
6. William Silverman, "The Economic and Social Effects of Automation in an Organization," *The American Behavioral Scientist* 9:10 (June 1966).

7. R. Blauner, *Alienation and Freedom: The Factory Worker and His Industry* (Chicago: University of Chicago Press, 1964).

8. Telecom Australia, *Background Papers—Seminar on Social Research and Telecommunications Planning* (Melbourne: Planning Directorate, August 1979), p. 57.

9. Hannah Arendt, *The Human Condition* (Garden City, N.Y.: Anchor Books, 1959).

7 Liability and Privacy Implications

The introduction of any innovative technology commonly offers the potential for new harm to individuals. Added costs to society often accompany promising benefits. In this chapter the liability and privacy implications of electronic message transfer are examined. First, the responsibilities of message service providers are reviewed to ascertain whether the legal system can handle new situations resulting from the introduction of electronic message transfer. The discussion pays special attention to the privacy implications of electronic message transfer because it is a sensitive area where the potential harm to society could be particularly significant.

In the next section the nature of possible problems that could result from the use of the new technology is explored. Then a legal framework is applied to providers of similar services to determine what responsibilities may befit electronic message service providers. Finally, the question is raised whether new legal rules are necessary to cope with liability situations that could develop from electronic message transfer. (Legal notions of liability are described in appendix 7A.)

Because much has been written concerning the concept of privacy, these ideas are not developed here. The subsequent discussion deals with three privacy issues that are particularly relevant in the context of electronic message-transfer developments. In each case, the nature of the potential problem is described and then various technical and social approaches to protecting privacy concerns are discussed.

Types of Potential Harm

It is commonly understood that few technological systems (or even people) operate totally accurately. The postal system is no exception. Table 7–1 summarizes a study of error rates presently occurring in conventional letter-delivery operations. In developing the hybrid system, RCA has recognized that completely error-free systems are difficult to build and usually prohibitively expensive. Table 7–2 compares recommended error rates for electronic message-transfer systems to tolerable error rates for conventional activities by type of error. If these recommendations are accepted by USPS, some isolated cases of individual harm could develop.

Table 7–1
Summary of Letter Error Rates[a]

Type of Letter	*Process Step*	*Any Error*	*Address Error*	*Text Error*	*Lost Letters*	*Stuffing Errors*
Bill/Statement	End-to-End	0.013485	0.013485	0.002925	0.00030	0.0004
	Bill Data Preparation	0.0116	0.0080	0.0029	0.0001	
	Data Processing	0.00005	0.000025	0.000025		
	Outgoing Mailroom	0.00056	0.00006		0.0001	0.0004
	USPS Delivery	0.00495	0.0048		0.00010	
Remittance (Payment)	End-to-End	0.051357	0.0098	0.006152	0.00040	0.035
	Remittance Letter Generation	0.0461	0.005	0.006	0.0001	0.035
	USPS Delivery	0.004905	0.0048		0.000105	
	Incoming Mailroom	0.0001			0.0001	
	Remittance Processing	0.000252		0.000152	0.0001	

Source: RCA Government Communications Systems, *Electronic Message Service*, Task AC, p. 5–13.

[a]*Letter error rate* is the ratio of letters with the indicated error to the ratio of all letters.

Table 7–2
Magnitude of Source Error Rates

Error Type	*Error Rate*	
	Electronic Message Transfer	*Conventional*
Address Errors	7×10^{-3}	1×10^{-2}
Text Errors	3×10^{-3}	5×10^{-3}
Lost Errors	1×10^{-4}	3×10^{-4}
Any Errors	8×10^{-3}	2×10^{-2}

Source: RCA Government Communications Systems, *Electronic Message Service,* Task Report AC, pp. 5–12, 5–14.

Generally, one can anticipate a variety of types of harm. For instance, message text or address inaccuracies could arise from transmission errors or encoding-decoding errors. Consequently, messages or information might be misrecorded, misdirected, or late, resulting in harm to the parties involved. Other types of harm include invasion of privacy or harm arising from destruction or mutilation of messages. All of these situations could result from machine error or by human actions (whether malicious or not). The issue of concern is who incurs liability in such situations and what options are available for preventing or compensating for resulting injuries?

To determine which duties (and hence liabilities) are likely to be assigned to the providers of an electronic message-transfer service, the discussion will rely on an analysis of appropriate antecedent technologies. By examining the laws applicable to computer activities that deal with data storage and manipulation and to conventional message service providers, such as USPS and telegraph and telephone companies, it is possible to suggest legal approaches to issues of liability.

Four types of harm will be investigated: the loss, delay, or mutilation of messages; alteration of wording or meaning; unauthorized disclosure; and physical injury. Reasoning by analogy is characteristically the most important legal technique for approaching new situations. Analogy, of course, implies that there is a difference as well as a similarity between the use of a new technology and its antecedents. A combination of logic and policy reasoning will be used to buttress the viability and legitimacy of reasoning from one category to the other.

The Duties of Message-Service Providers

Both telephone and telegraph services can be viewed as acceptable antecedent technologies from which to draw parallel legal analogies with electronic message transfer for several reasons. Both services approximate the role of electronic message-transfer services, but with important differences. End-to-end electronic message-transfer operations could resemble a telephone service where a terminal, electric circuitry, and a computer are provided, but where no direct participation by the company is necessary to transfer the message. Alternatively, electronic message-transfer hybrid operations could be similar to a telegraph service, in which a message is delivered to an operator who then mechanically transmits the message to a new destination and ultimately to the person for whom it is intended. In the second case, as with telegraph companies, an operator is involved in accepting, sending, receiving, and delivering the message. The duties and responsibilities of both the telephone and telegraph companies are similar in many ways. Clearly, a telegraph service has more numerous public duties (such as accuracy and confidentiality) than those imposed on telephone operators because it handles and transforms messages.

Our extensive familiarity with the telephone and telegraph render the legal duties of these services a matter of common sense. A telephone company must provide impartial service. It must not harm anyone (physically or otherwise). It may impose reasonable limits on its service. However, the simplicity of such legal responsibilities is deceptive. These duties can lead to very complicated issues of liability. For example, the failure of a telephone company to complete a call to the fire department that delays the arrival of firefighters could result in the loss of a building. This situation suggests a simple legal duty to give precedence to emergency calls, but this simple duty could cause an endless series of additional potential liabilities. The liability rules applicable to the providers of telephone, telegraph, computer, and postal services are reviewed below.

Telephone companies, as common carriers, must serve all customers who request service, impartially. Although companies may reasonably define and limit their obligations to provide adequate service (either geographically or technically), they must maintain sufficient resources to accommodate new subscribers. If a subscriber requesting service has complied with the reasonable requirements of a company, such as payment of past bills, the company cannot arbitrarily refuse to make a particular connection, in the absence of a bona fide objection to the particular call (such as an obscene call). There is also a duty to receive and transmit promptly. Generally, calls are received and transmitted in the order they come in, although there is an obligation to give emergency calls priority on party lines. Failure to receive or transmit a call within a reasonable time would be a breach of the public duty. In all of its activities, a telephone company has a duty to exercise the degree of care, diligence, and skill commensurate with its expertise and its undertaking.

A telephone company has an obligation to keep communications confidential. However, it is not liable for an unauthorized interception in the absence of their own negligence. An authorized disclosure of a message by a telephone company is presumed negligent and the burden is on the company to prove otherwise.

Finally, the telephone service has an obligation to avoid physically harming a person as a result of unsafe conditions (for example, in its building, personnel, facility, and equipment). For example, a telephone company is liable for the physical damage caused by a fallen telephone pole, a defective telephone receiver, or the actions of a telephone repair person where negligence is a contributing factor. The requirement of negligence would normally exclude liability for such acts of God as storms and other independently intervening forces.

Telegraph companies are responsible for the same common carrier duties required of telephone companies, including impartial and prompt service. They are also liable for negligent acts. But because they actually intercept and process messages, they have additional duties to the public. Each communication must be kept secret. This duty requires that companies exercise diligence in preventing unauthorized third-party intercepts and also mishandling of information within the company. Telegraph companies are held liable for unauthorized thefts, forgeries, and mutilation of messages by their employees by screening personnel to ensure against the unauthorized disclosure of information.

A message must be delivered to the person to whom it is sent. The sender must be promptly notified if there is any unreasonable delay in either message transmission or message delivery to the receiver. Finally, there is liability for errors in the transmission of a message. Any error in transmission is presumed to be the fault of the telegraph company and the burden of proof is on the company to prove otherwise. This duty also requires that the telegraph company make appropriate inquiries to clarify a message if the exact message is not clear.

For both telephone and telegraph companies there is, of course, a duty to fulfill any obligations arising from contracts to which they are parties.

Computer-service providers have an obligation to specify the services that are and are not provided.[1] In addition, to the degree that computers present new technical problems, such as piracy or falsification of documents, it is the obligation of companies to employ their full expertise in preventing these and all similar harms from befalling the customer. A failure to exercise this expertise diligently in protecting a consumer will almost certainly result in liability for the company.

USPS has significantly fewer legal duties than do retail businesses. Without a statute or contract, there is no liability for lost or damaged mail. Thus, first-, second-, third-, and fourth-class mails are deemed ''nonaccountable'' (that is, no tracing is performed or possible; no proof of delivery is given). Statutes cover the act of securing against losses. Although an old statute formerly limited indemnity to $300, present statutes now allow for unlimited coverage. Also, in the past, only the value of lost merchandise could be recovered. Current regulations specify that if the patron is willing to buy adequate insurance, subsequent losses may also be covered. For example, failure to deliver a contract that could result in the loss of future revenue for the sender can be claimed if the customer insured for that amount. In addition, there is liability for cash-on-delivery (COD) mail, registered mail, certified mail, and insured mail because of the specific contractual terms of the agreement to provide these services.

Negligent acts that cause nonmail-related harm (such as physical injury by a mail truck) are covered by the Federal Tort Claims Act, 39USC § 409 (c). Thus, USPS has a duty not to harm the public and is liable for a breach of this duty.

Liability Rules and Electronic Message Transfer

The heavy reliance on established law for emerging computer problems suggests that the basic obligations of businesses will not change because of their involvement in electronic message transfer. Under contract law, they will still be obliged to comply with the terms of their contracts. And, of course, they bear the usual responsibility not to breach their duty to another party, thereby avoiding a tort.[2] This principle was succinctly summarized by Roy N. Freed:

> Generally, the utilization of a computer, regardless of whether it is operated by the user or by someone else, does not create legal problems, but merely varies the severity of those problems that stem from other causes.[3]

Thus, it can be reasonably anticipated that although electronic message transfer may create new technical and social problems for the satisfaction of existing duties to the public and one's customers, it will not create new legal duties or liabilities.

The duties of those involved in telecommunications and electronic message transfer can be summarized simply. They must provide reasonably prompt, nondiscriminatory, accurate, and confidential services to the public. In providing service facilities, they may not harm persons or things either physically or mentally. The employment of computers and the advent of electronic message transfer will not alter any of these obligations.

What will change is the manner and care exercised to ensure compliance with these duties. When computers or electronic message transfer open up new ways of causing harm to customers, there is a clear obligation to exercise reasonable care, diligence, and expertise in eliminating this harm. In addition, the purveyor of electronic message transfer services must be careful to explain those services which are provided and those which are not. This exercise of care will avoid an unintentional warranty that certain results can be achieved when such a warranty is not intended.

The primary manner in which tort liability might arise in the electronic message-transfer area would be through the negligence of an employee. Negligence is the failure to exercise due diligence or care, a kind of fault. ''Strict liability'' (that is, liability without fault) does not apply to telephone or telegraph companies or to USPS. Therefore, before liability can be imposed for electronic message-transfer problems, some fault would need to be found.

In limited circumstances, primarily situations concerning the unauthorized disclosure of secret information, punitive damages have been imposed. Punitive-damage awards are above the amount of harm actually caused. They are designed to punish and deter similar conduct. Some states have made it a criminal offense to disclose the contents of a telegram. Similar provisions could be applied to the transmission of messages via electronic message transfer.

Finally, liability could arise from either an express or implied warranty. A warranty is a promise of indemnity against defects in an article or service sold. An assurance of an absence of defects can be both expressed and implied. Thus, if an electronic message-transfer service is supposed to perform certain functions and produce certain benefits for a business or individual, and if it fails to provide that service or benefit, a breach of a warrant may have occurred. Liability would be for the actual harm caused, as with a breach of a contract.

There are two possible means of limiting liability. One way would be to include a clause in a contract that expressly disclaims liability for negligent acts. However, past attempts by the telephone and telegraph industries to limit their liabilities in this manner have not been received favorably by the courts. The courts have noted the monopoly nature of these industries and the fact that telephone and telegraph companies are charged with a duty to the public that does not exist for the common retailer. Courts have generally concluded that it is unconscionable and a violation of public policy to allow a company contractually to disclaim liability where there is no real bargaining power on the part of the consumer. Such a precedent could have implications for electronic message transfer, particularly if USPS is the only provider of hybrid service.

A second means of avoiding liability is either through the rules and regulations of the particular public service commission or federal agency that regulates electronic message transfer (assuming there will be such a commission or agency) or by statute. It is likely that the courts would uphold a statutory elimination of liability where they might not uphold a contractual elimination of liability, because statutes reflect some measure of social consensus and have wider general applications. However, in circumstances where regulatory commissions have attempted to limit liability for negligence, courts have imposed liability for gross negligence (that is, conduct that is beyond mere negligence and connotes some willfullness on the part of the actor). Assuming that electronic message-transfer traffic will be interstate, federal law would govern liability and a state commission would not be able to regulate electronic message transfer (and hence to limit liability) unless specific authorization was obtained in Congress.

Security of Electronic Messages in Transfer

The privacy and confidentiality of messages will be an obvious concern to potential users of electronic message-transfer systems. To attract a sufficient user market, RCA has recommended that USPS provide electronic services that are perceived to be as secure as first-class letter service. But both the methods for violating message confidentiality and the means to protect it may differ somewhat between conventional and electronic services. For instance, conventional letters are protected from casual inspection by sealed envelopes. Furthermore, the security of conventional messages is protected by diligent monitoring of postal employees and by locked delivery and route mailboxes. Finally, a variety of postal statutes provide criminal sanctions for unlawful intrusions of both postal employees and private parties.

Both technical (such as locked boxes or employee monitoring) and legal deterrents may be applied to electronic services. For example, monitoring activities could be extended to those employees working with the electronic system. Public electronic mailboxes could be locked. And existing statutes could be extended to electronic services. However, such forms of protection would not be sufficient to safeguard the privacy of electronic messages.

A typical scenario of conspiracy suggests that electronic communications could be efficiently intercepted by using a computer to sift through volumes of messages, identifying only those concerned with specific topics or involving specific individuals or organizations. Other, more extreme scenarios depict situations in which messages are modified or deleted while still in electronic form or false or unauthorized messages are introduced. Clearly, the electronic nature of these messages makes them more susceptible to violations of privacy and security than conventional ones.

Although actual convictions for unauthorized disclosure or unlawful wiretaps are few, interceptions of electronic communications may be fairly prevalent. One-half of U.S. telephones may be vulnerable to tapping, and this equipment may eventually serve as a link to electronic message transfer systems.[5] In simple terms, the interception of electronic message transfer messages involves either tapping the wired subsystem, monitoring the satellite signals, or tampering with the system hardware or software. In all cases, the perpetrator must have adequate technical knowledge and possess the necessary equipment. In designing the hybrid system, RCA recommended that protection methods should foil the casual perpetrator while offering limited protection against more sophisticated eavesdroppers.[6]

A wide variety of equipment is commercially available to assist the determined eavesdropper. Table 7–3 lists characteristics of typical equipment used to intercept communications from wire or radio subsystems. There is a wide range of possibilities in terms of both cost and effectiveness. In general, digital satellite signals are fairly difficult to intercept because equipment must be near the up-link source or within the radiated down-link "footprint." Also, wires are typically easier to tap than cables

Table 7–3
Characteristics of Intercept Equipment

Intercept Equipment Component	*Availability*	*Estimated Cost*	*Effectiveness*
Terrestrial antenna (1 ea)	Commercial	$500–2,000	Moderate to high
Satellite antenna (1 ea)	Commercial	$20,000–600,000	Moderate to high
Receiver with demodulator	Commercial	$6,000–88,000	Moderate to high
Other terminating equipment	Self-made or commercial	$25–15,000	Low to high
Wire or Cable Intrusion			
Inductive tap	Commercial	Up to $60	Very high
Audio amplifier	Commercial	Up to $60	Very high
Headset	Commercial	Up to $60	Very high
Penetration tools (various)	Commercial	Up to $50	Not applicable
Optional equipment	Commercial	$25–$15,000	Low to high

Source: U.S. General Accounting Office, *Vulnerabilities of Telecommunications Systems to Unauthorized Use,* LCD–77–102 (Washington, D.C., March 31, 1979), pp. 10, 13.

because they are vulnerable to induction. Cable casings must be physically penetrated to permit monitoring.

Modification of system software is another means of revealing confidential messages. This method involves obtaining access to programming terminals and altering system instructions such that copies of messages can be delivered to the perpetrator. In all these methods of intrusion, messages can be copied, deleted, modified, or surreptitiously introduced.

There are various technical, social and legal approaches to protecting the security of messages. The technical alternatives basically involve either preventing system penetration or protecting the messages themselves. System penetration can be made more difficult by armoring cables, by substituting cables for wire, by employing testing and alarms to detect intrusion, and by physically isolating the equipment. The last method, which is really a mixed technical-social approach, could involve burying cables and wire, limiting access to land surrounding radio transceivers, restricting personal access to programming equipment, or limiting the programming capabilities of terminal users. A recent GAO survey reveals that most federal agencies concerned with data security (including USPS) emphasized protection of hardware and facilities rather than message protection techniques.[7]

The actual message can be protected in several ways. Digitized messages, such as those transmitted over electronic message-transfer systems, have a certain amount of inherent protection because the eavesdropper must have knowledge of the control protocols (such as the transmission rate or digital coding scheme). Also, high-capacity channels employing message switching are fairly secure from the unsophisticated perpetrator because individual messages must first be found among various message paths and then isolated from a large flow of data.

Cryptography is a more effective way to protect messages. The use of ciphers is not new in message communications. In the days of the colonial post, encryption was frequently used because of a widely acknowledged lack of message security.[9] Since those days, the art of cryptography has advanced to the point that relatively complex, single-cipher key techniques can be automatically applied to messages, using commercially available hardware such as data encryption standard (DES) microchips. However, some experts have criticized the message-security capabilities of single-cipher techniques like DES, because they feel that message security may be compromised by a need to restrict knowledge of a common message key only to the communicating parties and the possibility that such ciphers can be defeated by exhaustive computational efforts. As a result, some favor the use of public-key encryption.[10]

Public-key systems, which are still in a developmental stage, are slower than single-cipher systems. Such systems would require some organization to maintain a public record of encoding keys. A technology assessment of public-key cryptosystems by SRI suggests that if they become widely used, message security could be gained at a nominal cost.[11] Regardless of which cipher technique is chosen, decisions will have to be made concerning the proper implementation strategy (for example, whether to use link or end-to-end encryption).

The right to privacy for users of electronic message-transfer systems might be protected in various ways by the Constitution, common law, and existing statutes. Some of these legal rules protect the rights of individuals from government intrusion while others protect one private party from another. The Constitution provides limited protection for an individual's personal records and communications. Basically, the Fourth Amendment limits the power of government agents to enter private areas to make searches. Although it does not completely bar the seizure of private papers or communications, it does set forth certin procedural guidelines. However, this constitutional protection has been largely overshadowed by the promulgation of recent legislation.

In 1968, Congress enacted a comprehensive law covering the interception of wire or oral communications through wiretapping and electronic surveillance by government agents. This legislation provides substantial criminal and civil penalties for violations and details the procedures for obtaining authorizations. But loopholes in the statute have been used by government agents to justify electronic surveillance and wiretapping of citizens considered to be dangerous to national security. If such statutory interpretations are applied to electronic message transfer, users may be subject to authorized government surveillance. Past abuses of power by federal agencies have led to efforts, so far unsuccessful, to enact new legislation to restrict such abuses.

One general standing statute extends beyond limiting the actions of government agents. The Communications Act of 1934 states that one who receives or transmits a wire or radio communication shall not "divulge or publish the existence, contents, substance, purport, effect, or meaning thereof," except to the addressee or pursuant to court authorization. Under this law, violations are punishable by fine or imprisonment for up to 1 year. This law probably offers the most general legal protection for controlling future message security. However, communications may be divulged in accordance with federal wiretapping and electronic surveillance legislation.

Finally, common law may also provide some limited protection against interception of electronic message-transfer messages by conferring a right to collect damages. However, because this is an area of tort law, the plaintiff is expected to show actual harm. Also, tort law permits "reasonable" intrusions and publicity.

Statutes seem to provide the greatest legal protection for the security of electronic messages in transfer because they indicate social consensus and may be widely applied. If a legal approach to message security is taken, it will be necessary to test the ability of existing rules to curb any abuse of privacy associated with electronic message transfer. If these rules prove to be insufficient, new legislation may be necessary to prevent privacy abuses by government authorities or private individuals.

It is nearly impossible to guarantee the complete privacy of message communications. Any communication system that can be built can be tapped if the perpetrator has sufficient resources, knowledge, and time. Therefore, electronic message-transfer systems must be designed to make the costs of subversion much greater than its benefits to protect message privacy adequately. In this regard, electronic message-transfer developers might be encouraged to use a mix of technical and social approaches to make subversion costs unacceptable to all but the most determined eavesdropper. The government might also move to impose heavy penalties for surreptitious activities with respect to electronic message-transfer communications. These efforts may enhance the future social utility and attractiveness of electronic message-transfer services.

Data Security

The last section dealt chiefly with the security of messages in transfer. An important related issue concerns the effect of electronic message-transfer developments on the privacy of stored messages or data. A commission that recently studied the problem of privacy protection extensively drew the following conclusions:

> First, advances in computer and telecommunications technologies are dramatically and rapidly altering the way records about individuals are created, maintained, and used;
>
> Second, while computer and telecommunications technologies serve the interests of organizations and can be best appreciated as extensions of those interests, their broad availability and low cost provide both the impetus and the means to perform new record-keeping functions;
>
> Third, technology, like the law, has by and large failed to provide the tools an individual needs to protect himself from the undesirable consequences that recorded information can create for him today; and
>
> Fourth, growth in society's record-keeping capability threatens to upset existing power balances between individuals and organizations, and between government and the rest of society, thereby creating the danger that delay in addressing important privacy protection issues will irrevocably narrow the range of options open to public policy makers.[12]

Electronic message-transfer technology offers such possibilities.

Developments in electronic message-transfer may enhance the abilities of individuals or organizations to gather data on others for three reasons. First, the routine encoding of messages for transfer permits new opportunities for automated identification of items concerning specific topics or involving particular individuals. In addition to its automated sifting possibilities, encoding can reduce the expense and effort involved in accumulating data files. Finally, certain technical features of electronic message-transfer systems (such as the message-audit trail and testing and maintenance features) make it feasible to accumulate elaborate and traceable message records, including bank transactions, purchases, travel records, and correspondence between parties. As a result, electronic message-transfer developments could be considered as a threat to privacy, irrespective of whether data-gathering activities are licit or illegal, if they are not properly controlled.

To guard against such threats, service providers might be required to take reasonable precautions to avoid illicit system penetration and to restrict access to message audit records. In addition, legislation like the Privacy Act of 1974 might be necessary to control legal intrusions of privacy by private individuals. One particularly promising technical approach to controlling possible abuses is to cipher all electronic message-transfer communications so that they can be read only by intended recipients. Such efforts may be less costly than legal enforcement activities or other technical alternatives and provide sufficient protection against abuses of data security.

In summary, electronic message-transfer developments may further exacerbate the problem of reconciling conflicting social desires to both increase access to data and protect sensitive, personal information. And society may eventually face a tradeoff between increasing the security or the availability of electronic messages.

Intrusion

Electronic message-transfer systems could be a medium for conveying unwanted communications. There are two types of unwanted communications directed toward the home: offensive items and unwanted advertising or solicitations.[13] Because electronic message-transfer developments may decrease communication costs, many fear that the privacy of the home will be violated by a dramatic escalation of these types of messages. For instance, a particularly bleak scenario portrays a situation where an electronic message-transfer system user wades through a flood of solicitations deposited in his box or directed to his terminal to find a few personal messages. Such possibilities are not totally inconceivable, especially since RCA has projected that only one-third of the total anticipated traffic on a hybrid system might be correspondence.

There are legal, regulatory, and technical approaches to dealing with such possibilities. For instance, current postal regulations that allow individuals to stop delivery of any sexually oriented material or more general items from specific mailers could be extended to USPS electronic services. Also, standing government regulations and statutes (including FCC, public utility commission, state and local laws) covering the unlawful use of telephones for ''threatening, obscene, harassing, or annoying'' calls could be modified and extended to apply to electronic message-transfer services. Finally, the advertising industry itself could be encouraged to maintain an up-to-date listing of persons who wish to avoid receiving electronic mail solicitations.

Various technical alternatives are also possible. For instance, systems could be designed to block the delivery of messages from specific mailers or items dealing with particular topics automatically. Such sifting could be carried out at central processing centers or at programmed home terminals. Alternatively, home terminals could automatically list sender-supplied message headers that carry enough information to allow recipients to decide in which order to view messages or simply whether to view particular items at all.

It is clear that effective technical and legal approaches can control the potential problem of a massive intrusion of unwanted electronic messages in the privacy of the home. But a social conflict could arise between a sender's right of free speech and a receiver's right to avoid an intrusion of personal privacy, two contradictory social goals.

Responsibilities of Developers

Electronic message-transfer development could have a variety of significant consequences for liability and privacy. The introduction of these systems does not suggest that radically new forms of intrusions are likely or that new legal duties will be extraordinary. Rather, these systems will affect the magnitude and scale of harm. Because more information (in such forms as data, correspondence, and records) will be available in machine-readable form, it will be susceptible to interception. Various technical, legal, and social alternatives can adequately protect privacy. However, electronic message transfer may increase the possibility of lawful invasions of privacy even though a desire may exist to apply controls. Thus, new transfer technology may cause a confrontation between fundamental issues of national information and communication policy. For instance, individual rights of privacy may conflict with national security requirements. Such conflicts may raise major constitutional, legal, and social issues.

To fulfill legal duties reasonably and to avoid new liabilities, providers of electronic message-transfer service will have to take actions necessary to ensure that equipment does not malfunction with regularity (creating such problems as transmission errors, misrouting, or delayed information) and that the system or messages carried cannot be easily tampered with (allowing persons to modify or introduce bogus messages). Proper attention to system planning, equipment design, and security along with the use of precautionary measures such as encryption techniques may be costly; in the long run, however, the legal and social costs of neglecting such efforts may be even costlier. Although the legal process appears to have adequate means to deal with new situations of harm resulting from the introduction of electronic message transfer, unnecessary future costs to society might be avoided by implementing policies that require service providers to guard against such possibilities.

Notes

1. L.W. Smith, "A Survey of Current Legal Issues Arising from Contracts for Computer Goods and Services," *Computer/Law Journal* 1 (1979), pp. 475, 479.
2. See J.A. Philpott "Imposing Liability on Data Processing," *Santa Clara Law Review* 13 (1972), pp. 140, 148.
3. R.N. Freed, *Computers and Law* (Boston: R.N. Freed, 1976), p. 14.
4. RCA Government Communications Systems, *Electronic Message Service—System Definition and Evaluation* (Washington, D.C.: NTIS, 1978), executive summary, p. 43.
5. U.S. General Accounting Office, *Vulnerabilities of Telecommunications Systems to Unauthorized Use,* LCD–77–102 (Washington, D.C., March 31, 1979), p. 2. See also O.G. Selfride and R.T. Schwartz, "Telephone Technology and Privacy," *Technology Review* (May 1980).
6. RCA, *Electronic Message Service,* p. 46.
7. U.S. General Accounting Office, *Annotated Systems Security—Federal Agencies Should Strengthen Safeguards Over Personal and Other Sensitive Data,* LCD–78–123 (Washington, D.C., January 23, 1979).
8. See chapter 2 and appendix 2A for a discussion of techniques and technological components.
9. Wayne Fuller, *The American Mail* (Chicago: University of Chicago Press, 1972), p. 38.

10. D.E. Denning and P.J. Denning, ''Data Security,'' *ACM Computing Surveys* 2:3 (September 1979), p. 244.

11. SRI International, ''A Technology Assessment of Public Key Cryptosystems,'' Interim Presentation to National Science Foundation Division of Policy Research and Analysis, June 1980.

12. U.S. Privacy Commission, *Study* (Washington, D.C.: U.S. GPO, 1976), p. 1.

13. Walter S. Baer, ''Controlling Unwanted Communications to the Home,'' *Telecommunications Policy* (September 1978), pp. 218–228.

Appendix 7A Legal Notions of Liability

Liability is a legal status that involves a wrongful breach of duty by one party to another. A legal *duty* is an obligation by one party to another to act or not act in a particular manner that is recognized and enforced in law. Liability cannot be found if there is no legally recognized duty between the parties. Such a duty might arise because of a contract between the parties or by operation of the law of torts. A *contract* is an agreement between parties to do or refrain from doing some lawful thing. A tort is the commission or omission (by a person without right) of an act that results in an injury to the person, property, or reputation of another. *Tort* is a breach of a duty established by society generally, rather than by the parties themselves, as in the case of contract.

The societal expectation embodied in a tort duty may be expressed as a *statutory* or *regulatory* standard. The law may find a duty even in the absence of written standards or may impose a duty more stringent than written standards in particular circumstances. A *warranty* is part of a contract of sale, either expressed or implied, that affirms statements made during negotiations for a sale.

If liability cannot be avoided, then damages (usually money) will be awarded to the injured party as recompense for the injury. Damages can be assessed only for that harm which is ''proximately'' caused by the conduct of a party. In this situation, the consequences resulting from a breach of the duty must be the ordinary result of such a breach. The harm must be within the reasonable contemplation of the parties at the time of the contract or reasonably foreseen at the time of the tort. Furthermore, damages cannot be speculative.

Courts have generally required the injured party to mitigate the harm suffered. If mitigating steps have not been taken, the court will disallow any claim for that portion of the injury which occurred after the time that corrective steps should reasonably have been taken. The requirement of mitigation is merely another aspect of proximate cause.

Legal fault, a crucial concept for assigning liability, is not necessarily the same as *fault* in the moral sense. Societal expectations, and hence societal notions of fault, have changed over time with the evolution of the political economy and technological changes. Rules appropriate for the smooth operation of craft and cottage industries did not necessarily meet the needs of an emerging industrial society; a postindustrial society may evolve still different norms.

It appears that in the early development of common law in feudal England, one acted at one's peril, responsible for all the actual consequences of one's behavior. This situation is sometimes called "strict liability." Around the time of the Industrial Revolution, negligence became the basis for most tort liability. Because negligence requires only that one act reasonably, along generally accepted norms for one's position in common situations, the risk of acting for the actor became greatly reduced.

By the beginning of the twentieth century, mounting social pressures caused by industrial acccidents and by harm resulting from dangerous or poorly made goods forced the law to increase the number of situations in which strict liability would prevail. The theoretical justification for this evolution involved a notion that intentionally exposing others to risks for personal benefit should carry liability along with it. Strict liability in these more modern situations is clearly tied to notions of welfare economics consistent with a postindustrial society. Situations that have given rise to strict liability include new or strange activities (clearly dependent on types and rates of technological change), "abnormally dangerous" things or things used unnaturally or out of place, "ultrahazardous activities" (in which the risk cannot be eliminated even with utmost care), and a hard-to-define legalism (or conclusion) of "absolute nuisance," which also refers to inappropriate usage. One incurs no liability for accidents that are unavoidable or that result from "normal" conduct common to all persons or corporations of the same type.

The notion of fault with regard to liability for a modern technological development such as electronic message transfer can be seen as a social concept in which various factors can be brought to bear in balancing the social utility of the conduct in question against the probability and gravity of risks created by it. These factors include:

Social judgment on the norms required for society for the protection of others (only sometimes related to moral notions of fault)

Historical practice and traditions

Convenience of administration, not unnecessarily adding to the friction of life in an already complex society

Capacity to bear loss (internalization of externalities or equity or distributional justice; colloquially known as the "deep-pocket" theory)

Capacity to socialize loss, or spread it among large numbers of people (each of whom carries only a small amount, as when the price of a consumer good includes the proportionate fraction of industrial health and safety provisions)

Capacity to avoid loss, as a spurt to conduct that would prevent the loss in the first place

Punishment, in situations where the conduct has a reprehensible element (sometimes sanctions are also seen as deterrence to others).

8 Electronic Message-Transfer Developments in Perspective

In the preceding chapters, it has been suggested that publicly available electronic message-transfer services will be far more widespread in the next decade. The technology necessary for viable electronic message-transfer systems is already in place. New technological developments will further reduce system costs. Correspondingly, the market potential for such services will continue to increase.

At present, because of relatively high costs for remote terminal and telecommunication linkages, end-to-end service will be employed primarily by heavy users. In the near term, the general public can be expected to use hybrid systems that use both conventional and electronic technology. Because terminal and linkage costs will certainly fall, hybrid-service users may switch to end-to-end services over the longer term.

The development of widely used end-to-end electronic transfer systems will be more feasible when the general public acquires terminal equipment. Decisions concerning terminal acquisitions will depend not only on cost factors but also on multiple-use-capabilities. If remote terminals can be used for a number of desirable functions, users may justify acquisition on a basis of shared costs. Such applications may include payment of accounts and bookkeeping, retention of records (such as recipes and family birthdays), home computing, text editing, information sorting and retrieval, information dissemination (such as news or stock prices), home purchases (for example, from sales catalogs), entertainment, and teleprocessing.

The advent of personal computers, videotext, teletext, and two-way cable systems might help to establish a mass computer market for electronic message systems. The popularization of these systems will drive terminal costs down and allow users to discover the utility of acquiring such equipment for information storage and retrieval, message transfer, news and information, or entertainment.

As a result, users will switch from conventional message transfer systems to electronic alternatives for some communications needs because of its perceived advantages (such as cost, speed, convenience, or novelty). Others will discover new uses. As a result, the volume of electronic messages can be expected to grow rapidly unless consumer reaction is unfavorable, more attractive alternatives merge, or other previously mentioned factors inhibit growth. The nature and composition of electronic message traffic will also depend on the character of service providers and on government policies and requirements in regional and national markets.

Much of the present conventional mail could be diverted to an electronic system. However, without special government incentives to develop a mass market, all-electronic services may be initially designed to meet the needs of the business community. An initial concentration on the business market for all-electronic services is similar to the early diffusion pattern of telephone services.

Electronic message-transfer services can offer capabilities that combine several features of telephone and mail services. Electronic systems can be as interactive and rapid as telephone systems. But, like the mail, these systems can produce written records of the communications. The development of a mass market for electronic services may depend upon the typing skills of the public unless low-cost facsimile services emerge or voice transcription methods become feasible.

Nationwide electronic message-transfer systems will require a very large capital expenditure. Yet, if conventional transportation, processing, and handling message costs are compared the costs of parallel electronic services, the electronic services are more cost efficient because of the higher labor costs associated with conventional processing techniques.

If the national postal monopoly on conventional mail is extended to electronic services, speed of innovation and costs are likely to suffer. The entry of private firms into previously restricted telecommunications markets has spurred the pace and adoption of innovations and reduced costs. Generally, competition in the end-to-end market may benefit consumers. However, sole-service providers might be justified in specific regional markets where demand is limited.

If USPS is not excluded from developing an electronic service, its involvement in early system developments might stimulate industrial activity and inspire public confidence in the viability of the new service. Furthermore, government involvement could ensure that services will be developed in all markets, not just the most lucrative. Thus, joint ventures might be valuable arrangements. Few private organizations have the infrastructure readily available to USPS. The combination of USPS resources and expertise available in the private sector may help to make hybrid electronic systems, in particular, successful.

It may be desirable to maintain two written message-transfer alternatives to provide increased consumer choice and also for national security. Public opinion, at least in the short run, is likely to demand that electronic services coexist with conventional services. People may continue to appreciate conventional transfer of intrinsically valued messages such as perfumed stationary or greeting cards. A situation may evolve in which electronic services are used for rapid communication while conventional mail services are used to transfer more personal, intrinsically valued items.

If electronic message-transfer services are widely used by the public and if USPS is involved in such developments, USPS will have to change the nature of its operations, alter its capital-labor mix, and attract more technically trained personnel. USPS will also have to develop a new public image and promote different customer usage patterns. If USPS-provided electronic transfer services are popular, the Postal Service may begin to cope with long-standing operational and deficit problems. If USPS is prohibited from offering electronic message transfer, its viability may be at stake.

Electronic message-transfer developments will affect postal and office labor. Such developments will change skill level requirements of employees. Projected increases in labor productivity may be accompanied by a reduced number of entry-level positions (for example, those requiring only secondary education) in both post and office industries. Some job dislocations are likely. Electronic message-transfer developments can also enhance work conditions and employee opportunities. Government and industry cooperation could augment these possibilities.

The duties and responsibilities of the providers of electronic services would not radically differ from those of conventional service providers. While providers of new services may have few additional liabilities, the scale of harm may be larger as electronic message volumes increase. However, current legal rules can probably cope with these situations of liability.

Electronic message-transfer developments may have new implications for privacy, especially in the areas of security, data protection for files and records, and the intrusion of communications. A mix of technical, legal, and social alternatives might be used to cope with potential problems effectively and efficiently. However, society's (or government's) right to know may be in conflict with an individual's right of privacy.

Electronic message-transfer may be viewed as a depersonalizing medium, perhaps promoting social alienation. The infusion of rapid communications and information processing into daily life may raise questions regarding information overload and personal autonomy in decision making. Electronic message services may change the nature of the human communication activity. Users may expect faster responses to messages. Creditors may demand more timely payment of bills. Abbreviated notes that are expeditiously communicated may lack subtlety and nuance. Consumers must learn to use the technology carefully and wisely.

Finally, the use of electronic message services may conserve some scarce resources, particularly energy, paper, and vehicle requirements for postal operations. Electronic message-transfer developments could also contribute some of the necessary infrastructure required to decentralize office work locations, encourage teleshopping, and provide work-at-home alternatives. These possibilities could have profound impacts upon national energy consumption, methods of sales, inventory control, personnel management, and work and travel patterns in the white-collar and retailing industries.

It is difficult to conclude whether electronic message-tranfer systems will yield net benefit or harm to society. Even if such systems are developed and deployed in a well-planned manner there will still be both costs and benefits. Many of these anticipated effects cannot be quantified. This, it is difficult to conclude whether the benefits of electronic message transfer will outweigh its costs. However, the benefits that could result are not trivial, especially productivity gains in the office and postal workplace, resource conservation, and balance of trade. The social costs can be mitigated by formulating policies and enacting laws that cope with labor, liability, and privacy issues. Furthermore, the magnitude of the costs and benefits of electronic systems will be conditioned by the selection of implementation policies. If the developers promote the potential benefits of electronic message transfer systems and encourage efforts to alleviate costs, such technological possibilities can be designed and implemented is a socially beneficial manner.

Bibliography

Adie, Douglas K. *An Evaluation of Postal Service Wage Rates* Washington, D.C.: American Enterprise Institute for Public Policy Research, 1977.

———. "Has the 1970 Act Been Fair to Workers?" (Paper prepared for a Conference on Postal Service Issues, October 1978).

Administrative Management Society. *AMS Professional Management Bulletin* 14, June 1976.

Air Transport Association of America. *Air Transport*. Washington, D.C., 1977.

American Enterprise Institute. *Postal Service Legislative Proposals*. Washington, D.C.: American Enterprise Institute for Public Policy Research, 1977.

APEX Word Processing Working Party. *Office Technology—The Trade Union Response*. London:. APEX, March 1979.

Arendt, Hannah. *The Human Condition*. Garden City, N.Y.: Anchor Books, 1959.

Arthur D. Little Inc. *Electro-Optic Technology in Local (EMS) Terrestrial Networks*. Washington, D.C.: NTIS, 1974.

———. *The Consequences of Electronic Funds Transfer*. Washington, D.C.: U.S. GPO, 1975.

———. *Telecommunications and Society, 1976–1991,* Report to the Office of Telecommunications Policy, Executive Office of the President. Washington, D.C.: NTIS, 1976.

———. *The Application of Contextual Analysis in Message Body Optical Character Recognition*. Washington, D.C.: NTIS, 1976.

———. *A Survey of Factors Affecting the Quality of Data Compressed Black/White Facsimile*. Washington, D.C.: NTIS, 1976.

——. *Paper and Paper Substitutes for Electronic Message Systems*. Washington, D.C.: NTIS, 1976.

Baer, Walter S. "Controlling Unwanted Communications to the Home," *Telecommunications Policy,* September 1978, pp. 218–228.

———. *Telecommunications Technology in the 1980's*. Santa Monica, Cal.: (Rand Paper P–6275), December 1978.

Baran, Paul. *Potential Market Demand for Two-Way Information Services to the Home*. Menlo Park, Cal.: Institute for the Future, December 1971.

Baratz, Morton S. *The Economics of the Postal Service*. Washington, D.C.: Public Affairs Press, 1962.

Bell, Daniel. *The Coming of Post-Industrial Society*. New York: Basic Books, 1973.

Blanc, Robert P. *Cost Analysis for Computer Communications*. Washington, D.C.: National Bureau of Standards, 1974.

Blau, Peter, and Otis Dudley Duncan. *American Occupational Structure*. New York: Wiley, 1967.

Blauner, R. *Alienation and Freedom: The Factory Worker and His Industry*. Chicago: University of Chicago Press, 1964.

Bok, Derek. *Automation, Productivity and Manpower Problems*. Washington, D.C.: U.S. Department of Labor, 1964.

Booz Allen. ''Office Automation in the '80s'' (Mimeo of seminar presentation), 1980.

Braverman, Harry. *Labor and Monopoly Capital*. New York: Monthly Review Press, 1974.

Brock, G.W. *The Telecommunications Industry*. Cambridge, Mass.: Harvard University Press, 1981.

Bowsers, R., A.M. Lee, and C. Hershey, ed. *Communications for a Mobile Society—An Assessment of New Technology*. Beverly Hills, Cal.: Sage, 1978.

Campbell, J.I. ''Politics and the Future of Postal Services'' (paper prepared for a Conference on Postal Service Issues, October 13, 1978).

Commission on Postal Service. *Report*. Washington, D.C.: U.S. GPO, 1977, vols. 1, 2, and 3.

Committee on Planning and Design Policies. *Road User Benefit Analyses for Highway Improvements*. Washington, D.C.: American Association of State Highway Officials, 1960.

Coven, Robert, ''New Cryptography to Protect Computer Data,'' *Technology Review* December 1977, pp. 6–7.

Crozier, Michael. *The Bureaucratic Phenomenon*. Chicago: University of Chicago Press, 1967.

Cullinan, Gerald. *The United States Postal Service*. New York: Praeger, 1974.

Denning, D.E., and P.J. Denning. ''Data Security,'' *ACM Computing Surveys* 2 (3), September 1979.

Department of Communications, Canada. ''Postal Services and Telecommunications,'' study (7)i. Ottawa: Information Canada, 1972.

Dickson, E.M., and R. Bowers, *The Video Telephone, Impact of a New Era in Telecommunications*. New York: Praeger, 1974.

Doll, D.R. ''Multiplexing and Concentration,'' *Proceedings of the IEEE* 60 (11), November 1972.

Dunlop, J.T. *Automation and Technical Change*. Englewood Cliffs, N.J.: Prentice-Hall, 1962.

Engel, G.H., and others. ''An Office Communications System,'' *IBM Systems Journal* (18) 3, 1975, pp. 402–431.

External Telecommunications Executive. *Preliminary Report on Word Processors and International Communications*. Long Range Studies Division, British Post Office, September 1978.

Ferguson, C.E. *Microeconomic Theory*. Homewood, Ill.: Richard D. Irwin, 1973.

Ford, K.W. "The Characteristics of Automatic Editing Typewriters," *The Office,* February 1976, pp. 63–67.

Fowler, D.G. *Unmailable*. Athens: University of Georgia Press, 1977.

Freed, R.N. *Computers and Law*. Boston: R.N. Freed, 1976.

Frey, J., and A. Lee. "Technology of Land Mobile Communications," in *Communications for a Mobile Society—An Assessment of New Technology,* ed. R. Bowers, A. Lee, C. Hershey, Beverly Hills, Calif.: Sage, 1978.

Fuller, Wayne. *The American Mail*. Chicago: University of Chicago Press, 1972.

Fuss, Melvyn. "Cost Allocation: How Can the Costs of Postal Service be Determined?" (paper prepared for a Conference on Postal Service Issues, October 13, 1978).

Geller, Henry, and Stuart Brotman. "Electronic Alternatives to Postal Service," in *Communications for Tomorrow,* ed. Glen O. Robinson. New York: Praeger, 1978.

Gellman Research Associates. *Economic Regulation and Technological Innovation: A Cross National Literature Survey and Analysis*. Washington, D.C.: NTIS, 1974.

General Dynamics/Electronics. *Study of Electronic Handling of Mail*. Washington, D.C.: NTIS, 1970.

George, James E. "Word Processing—Present and Future," *Proceedings of AESOP Conference*. Seattle, Wash., May 3–4, 1977.

Georgi, H. *Cost-Benefit Analysis and Public Investment in Transport*. London: Butterworth, 1973.

Ginzburg, Eli. "The Professionalization of the U.S. Labor Force," *Scientific American* 240 (3), March 1979.

Glenn, E.N., and R.L. Feldberg. "Degraded and Deskilled: The Proletarianization of Clerical Work," *Social Problems,* October 1977, pp. 52–64.

Goldman, T.A., ed. *Cost-Effectiveness Analysis,* New York: Praeger, 1967.

Graham, William M. "Direct Testimony on Behalf of the U.S. Postal Service," unpublished manuscript, 1977.

Gray, P. *Prospects and Realities of the Telecommunictions/Transportation Tradeoff*. Los Angeles: University of Southern California, Center for Futures Research, November 1974.

Haldi, J., and J.F. Johnston. *Postal Monopoly: An Assessment of the Private Express Statutes*. Washington, D.C.: American Enterprise Institute for Public Policy Research, 1974.

Hall, Arthur D. *The Economies of Scale in Today's Telecommunictions Systems*. New York: IEEE Press, September 1973.

Hall, Richard. *Occupation and Social Structure*. Englewood Cliffs, N.J.: Prentice-Hall, 1975.

Harkness, R.C. "Telecommunications Substitutes for Travel, A Preliminary Assessment of Their Potential for Reducing Urban Transportation Costs by Altering Office Location Pattern (unpublished Ph.D. Dissertation), University of Washington, Seattle, 1973.

———. *Technology Assessment of Telecommunications/Transportation Interactions*. Washington, D.C.: NTIS, 1977.

———. "Office Information Systems," *Telecommunications Policy,* June 1978, pp. 91–105.

Hellman, Martin. "The Mathematics of Public-Key Cryptography," *Scientific American,* August 1979, pp. 146–157.

Hiltz, S.R., and Murray Turoff. "EFT and Social Stratification in the USA," *Telecommunications Policy,* March 1978, pp. 22–32.

Holsendolph, E. "U.S. Postal Service Plans to Start Electronic Mail Service This Year," *The New York Times,* February 14, 1978, p. 1.

———. "U.S. Postal Service and Comsat to Test Electronic Mailings," *The New York Times,* March 29, 1978, p. 1.

———. "House, Unhappy with Postal Service, Seeks Changes," *The New York Times,* April 17, 1978, p. 13.

Institute for the Future. "Technology Assessment of Teletext and Videotex in the U.S.," NSF Grant PRA 80–12731.

Institute for Telecommunication Sciences. *Study of Satellite Frequency Requirements for the U.S. Postal Service Electronic Mail System*. Washington, D.C.: NTIS, 1973.

———. *A Digital Telecommunications Conversion Model for Electronic Message Mixtures*. Washington, D.C.: NTIS, 1975

———. *Analysis of Digital Telecommunications Message Model*. Washington, D.C.: NTIS, 1975.

———. *A Numerical Method for Generating Earth Coverage Footprints in Geostationary Antennas*. Washington, D.C.: NTIS, 1975.

———. *Study of Error Control Coding for the USPS Electronic Message System*. Washington, D.C.: NTIS, 1975.

———. *Network Capacity and Queue Aspects of USPS Electronic Message Systems*. Washington, D.C.: NTIS, 1976.

International Resource Development. "Videoprint," *Newsletter* 1, 1980.

Jackson, C.L. *Electronic Mail,* Report CSR TR–73–2 NASA Contract no. 2197, April 1973.

———. "What Will New Technology Bring?" (paper prepared for a Conference on Postal Service Issues, October 13, 1978).

Kahn, A.E. *The Economics of Regulation.* New York: Wiley, 1971.

Kahn, A.E. "Applying Economics to an Imperfect World," *Regulation,* November–December 1978, pp. 17–27.

Kalba Bowen Associates. *Electronic Message Systems: The Technological Market and Regulatory Prospects,* submitted to the FCC, April 1978.

Kallick, M., W. Rodgers, and others. *Household Mailstream Study Final Report,* prepared for Mail Classification Research Division, USPS, 1978.

Karydes, Marianne, and others. *An Analysis of Domestic Public Message Telegraph Service.* Washington, D.C.: NTIS, 1973.

Kluttz, James E. "Workpapers" (unpublished manuscript), 1977.

Kuehn, R.A. *Cost-Effective Telecommunications,* New York: AMACOM, 1975.

Kulp, G., and others. *Transportation Energy Conservation Data Book,* 4th ed. Washington, D.C.: NTIS, 1980.

Lee, A.M., and P.L. Bereano. "Developing Technology Assessment Methodology: Some Insights and Reflections," *Technology Forecasting and Social Change* 19, 1981, pp. 15–31.

Lee, A.M., and A.H. Meyburg. "Some Policy Issues in Electronic Message Transfer," *Proceedings,* 1980 Annual Conference of the American Society of Engineering Education, vol. 2, 1980, pp. 292–295.

———. "Technology Assessment," in *Engineering: Cornell Quarterly* 15 (4), Spring 1981, pp. 7–15.

———. "Electronic Message Transfer in the Letter Post Industry: Resource Implications," in *Transportation Research Record* 812, National Research Council, 1981, pp. 59–62.

Lee, Ronald B. "The U.S. Postal Service," in *Urban Commodity Flow,* Special Report 102. Washington, D.C.: Highway Research Board, 1971.

Lipoff, S.J. "Mass Market Potential for Home Terminals," *IEEE Transactions on Consumer Electronics,* May 1979.

Mail Classification Research Division, USPS. "Phase III Study Plan," unpublished draft, 1975.

Mail Classification Research Division, USPS. "Feasibility of Obtaining a Quantitative Description of Non-Household Mailstream," unpublished draft, 1975.

Mail Classification Research Division, USPS. ''Feasibility of Obtaining a Quantitative Description of the Household Mailstream,'' unpublished draft, 1975.

Mail Classification Research Division, USPS. ''Service-Related Products Study,'' unpublished draft, 1977.

Martin, James. *The Wired Society*. Englewood Cliffs, N.J.: Prentice-Hall, 1978.

Mayo, R.W., and W.W. Wittman. *The Structure, Conduct, and Performance of the United States Telecommunications Industry*. Washington, D.C.: NTIS, 1977.

McBride, Charles C. ''Post Office Mail Processing Operations,'' in *Analysis of Public Systems,* ed. Alvin W. Drake. Cambridge, Mass.: MIT Press, 1974.

McNamara, J.E. *Technical Aspects of Data Communication*. Maynard, Mass.: Digital Equipment Corporation, 1978.

Memmott, F.W., III. ''The Substitutability of Communications for Transportation,'' *Traffic Engineering,* February 1963, pp. 20–25.

Merewitz, L. ''Costs and Returns to Scale in U.S. Post Offices,'' *Journal of the American Statistical Association* 66 (335), September 1971, pp. 504–509.

Messer Associates, Inc. ''Final Report: Functional Mail Flow Study,'' unpublished manuscript, 1977.

Meyburg, A.H., and R.H. Thatcher. ''The Use of Mobile Communications in the Trucking Industry,'' *Transportation Research Record 668*. Washington, D.C., 1979, pp. 21–22.

———. ''Land Mobile Communications and the Transportation Sector: Passenger Transportation,'' chap. 7 in *Communications for a Mobile Society—An Assessment of New Technology,* ed. R. Bowers, A.M. Lee, C. Hershey. Beverly Hills, Calif.: Sage, 1978.

Meyer, Martin. ''The Telephone and the Uses of Time,'' in *The Social Impact of the Telephone,* ed. I. Pool. Cambridge, Mass.: MIT Press, 1977, pp. 225–245.

Miller, F.W. ''Electronic Mail Comes of Age,'' *Infosystems* 24, November 1977, pp. 58–64.

Miller, James C., and Roger Sherman. ''Has the 1970 Act Been Fair to Mailers?'' (paper prepared for Conference on Postal Service Issues, October 13, 1978).

Mishan, E.J. *Cost-Benefit Analysis*. New York: Praeger, 1976.

Mitre Corporation. *The Impact of Telecommunication on Transportation Demand Through the Year 2000*. Washington, D.C.: NTIS, November 1978.

Musgrave, R.A., and P.B. Musgrave. *Public Finance in Theory and Practice*. New York: McGraw-Hill, 1973.

Mustafa, Husain, M. *The Mechanization and Automation of the United*

States Post Office. Washington, D.C.: Center for Technology and Administration, 1964.

Myers, Robert J. *The Coming Collapse of the Post Office*. Englewood Cliffs, N.J.: Prentice-Hall, 1975.

National Academy of Sciences. *Technology: Processes of Assessment and Choice,* Washington, D.C.: U.S. GPO, 1969.

National Association of Letter Carriers. "The Electronic Mail Issue: The Issues, the Players, and NALC's Position," mimeo, 1979.

National Bureau of Standards. *Methods for Synthesizing Networks of Electronic Mail Systems*. Washington, D.C.: NTIS, 1974.

———. *Mathematical Methods of Site Selection for Electronic Message Systems*. Washington, D.C.: NTIS, 1975.

National Commission on Technology. *Automation and Economic Progress, Technology and the American Economy*. Washington, D.C.: U.S. GPO, 1966.

National Research Council. *Study of Practical Applications of Space Systems—Uses of Communication*. Washington, D.C.: NTIS, 1974.

———. *Electronic Message Systems for the U.S. Postal Service,* Washington, D.C.: NTIS, 1976.

——. *Telecommunications for Metropolitan Areas: Near-Term Needs and Opportunities*. Washington, D.C.: NTIS, 1977.

———. *Review of Electronic Mail Service Systems Planning for the U.S. Postal Service*. Washington, D.C.: National Academy Press, 1981.

Navy Electronics Laboratory Center. *First Annual Report: Advanced Mail Systems Scanner Technology*. October 22, 1975.

Nilles, J.M., and others. *Telecommunications Substitutes for Urban Transportation*. Los Angeles: University of Southern California, Center for Futures Research, November 1974.

Nyborg, Philip S. "Regulatory Inhibitions on the Development of Electronic Message Systems in the USA," *Telecommunications Policy* 2 (4), December 1978, pp. 316–326.

Oettinger, Anthony, and others. *High and Low Politics: Information Resources for the 80's*. Cambridge, Mass.: Ballinger, 1977.

Panko, Raymond. "The Outlook for Computer Mail," *Telecommunications Policy,* June, 1977.

Parker, Donn B., Susan Nycum, and others. *Computer Abuse*. Menlo Park, Calif. Stanford Research Institute, November 1973.

Peat, Marwick, Mitchell, and others. *Industrial Energy Studies of Growth Freight Transportation*. Washington, D.C.: NTIS, July 1974.

Pempel, T.J. "Land Mobile Communications in Japan: Technical Developments and Issues of International Trade," in *Communications for a Mobile Society,* ed. R. Bowers, A. Lee, C. Hershey, Beverly Hills: Sage, 1978, pp. 317–343.

Philco-Ford Corporation. *Conversion Subsystems for the Electronic Mail Handling Program*. Washington, D.C.: NTIS, 1973.

Phillips, C.F. *The Economics of Regulation*. Homewood, Ill.: Irwin, 1969.

Philpott, J.A. ''Imposing Liability on Data Processing,'' *Santa Clara Law Review* 13, 1972.

Piore, M.J. ''Notes for a Theory of Labor Market Stratification,'' in *Labor Market Segmentation,* ed. R.C. Edwards and others. Lexington, Mass.: D.C. Heath, 1981.

Pool, Ithiel de Sola, ed. *The Social Impact of the Telephone*. Cambridge, Mass.: MIT Press, 1977.

Porat, Marc. *The Information Economy: Definition and Measurement,* U.S. Department of Commerce, Office of Telecommunications, Special Publication 77–12(1), May 1977.

Post Office Engineering Union. *The Modernization of Telecommunications*. London: College Hill Press, June 1979.

Potter, R.J. ''Electronic Mail,'' *Science* 195, March 18, 1977.

President's Commission on Postal Organization. *Towards Postal Excellence*. Washington, D.C.: U.S. GPO, 1968.

Priest, George. ''The History of the Postal Monopoly in the United States,'' *Journal of Law and Economics,* April 1975, pp. 33–80.

RCA Government Communications Systems. *Electronic Message Service—System Definition and Evaluation*. Washington, D.C.: NTIS, 1978.

Revenue and Cost Analysis Division, USPS. ''Summary Description of USPS Development of Costs by Segments,'' July 1977.

Robinson, Howard. *Britain's Post Office*. London: Oxford University Press, 1953.

Robinson, Kenneth. ''The Postal Service and Electronic Communications: Legal Issues and Open Questions'' (paper prepared for a Conference on Postal Service Issues, October 13, 1978).

Rosch, Gary. ''Viewdata and Teletext Systems: What the Europeans are Doing'' (paper prepared for Teletext and Videotext in the U.S., Workshop on Emerging Issues, Institute for the Future, Pajaro Dunes, Calif., June 20–22, 1979).

Scheele, Carl. *A Short History of the Mail Service*. Washington, D.C.: Smithsonian Institution Press, 1970.

Selfride, O.G., and R.T. Schwartz. ''Telephone Technology and Privacy,'' *Technology Review,* May 1980.

Shepard, Jon M. *Automation and Alienation: A Study of Office and Factory Workers*. Cambridge, Mass.: MIT Press, 1971.

Sherman, Roger, ed. *Perspectives on Postal Service Issues*. Washington, D.C.: American Enterprise Institute, 1980.

Shonka, D.B., ed. *Transportation Energy Conservation Data Book,* 3d ed. Washington, D.C.: NTIS, 1979.

Silverman, William. ''The Economic and Social Effects of Automation in an Organization,'' *The American Behavioral Scientist* 9 (10), June 1966.

Singer, N.M. *Public Microeconomics*. Boston: Little, Brown, 1972.

Sirbu, M.A. ''Innovation Strategies in the Electronic Mail Marketplace,'' *Telecommunications Policy,* September 1978, pp. 191–210.

———. ''Automating Office Communications, Policy Dilemmas,'' *Technology Review,* October 1978, pp. 50–57.

Smith, L.W. ''A Survey of Current Legal Issues Arising from Contracts for Computer Goods and Services,'' *Computer/Law Journal* 1, 1979.

Sorkin, Alan. *The Economics of the Postal System*. Lexington, Mass.: Lexington Books, D.C. Heath, 1980.

SRI International. ''A Technology Assessment of Public Key Cryptosystems,'' interim presentation, June 1980.

Stevenson, Rodney, E. ''The Pricing of Postal Services,'' in *New Dimensions in Public Utility Pricing,* ed. Harry M. Trebling. MSU Public Utilities Studies, East Lansing, Mich., 1976.

Telecom Australia. *Background Papers—Seminar on Social Research and Telecommunications Planning*. Melbourne: Planning Directorate, August 1979.

Transportation Association of America. *Transportation Facts and Trends*. Washington, D.C.: Transportation Association of America, 1977.

Transportation Research Board. ''Cost-Benefit and Other Economic Analyses of Transportation,'' *Transportation Research Record 490,* NRC, Washington, D.C., 1974.

Tyler, Michael. ''Electronic Publishing: A Sketch of the European Experience'' (paper prepared for Teletext and Videotex in the U.S., A Workshop on Emerging Issues, Institute for the Future, Pajaro Dunes, Calif. June 20–22, 1979).

U.S. Congress, House Subcommittee on Postal Operations and Services and the Subcommittee on Postal Personnel and Modernization of the Committee on Post Office and Civil Service. *Joint Hearings on HR 7700,* 95th Congress, 1st Session. Washington, D.C.: U.S. GPO, 1977.

U.S. Department of Commerce, Bureau of the Census. *Historical Statistics of the United States*. Washington, D.C.: U.S. GPO, 1975.

———. *Statistical Abstract of the United States*. Washington, D.C.: U.S. GPO, 1978.

U.S. Department of Commerce, Office of Telecommunications. *Telecommunications Substitutability for Travel: An Energy Conservation Potential*. Washington, D.C.: U.S. GPO, 1975.

———. *The Contribution of Telecommunications to the Conservation of Energy Resources*. Washington, D.C.: NTIS, 1977.

———. *The Postal Crisis: The Postal Function as a Communications Service*. Washington, D.C.: U.S. GPO, 1977.

U.S. Department of Justice. *Changing the Private Express Laws*. Washington, D.C.: U.S. GPO, 1977.

U.S. Department of Labor. *Public Policy and Economic Dislocation of Employees,* Office of Assistant Secretary for Policy Evaluation and Research. Washington, D.C., October 1978.

U.S. General Accounting Office. *Missent Mail—A Contributing Factor to Mail Delay and Increased Costs,* Report to Congress. Washington, D.C., 1974.

———. *$100 Million Could be Saved Annually in Postal Operations Without Affecting the Quality of Service* GGD 75–87. Washington, D.C., 1975.

———. *A Summary of Observations on Postal Service Operations from July 1971 to January 1976* GGD 76–61. Washington, D.C., 1976.

———. *Challenges of Protecting Personal Information in an Expanding Federal Computer Network Environment* LCD–76–102. Washington, D.C., April 12, 1978.

———. *Annotated Systems Security—Federal Agencies Should Strengthen Safeguards over Personal and Other Sensitive Data* LCD–78–123. Washington, D.C., January 23, 1979.

———. *Vulnerabilities of Telecommunications Systems to Unauthorized Use* LCD–77–102. Washington, D.C., March 31, 1979.

———. *Developing a Domestic Common Carrier Telecommunications Policy: What Are the Issues* CED–79–18. Washington, D.C., January 24, 1979.

———. *Comparative Growth in Compensation for Postal and Other Federal Employees Since 1970* FPCD–78–43. Washington, D.C., February 1, 1979.

———. *United States–Japan Trade: Issues and Problems* ID–79–53. Washington, D.C., September 21, 1979.

———. *A Case Study of Why Some Postal Rate Commission Decisions Took As Long As They Did* GED–81–96. Washington, D.C., September 8, 1981.

———. *Implications of Electronic Mail for the Postal Service's Work Force* GGD–81–30. Washington, D.C., February 6, 1981.

U.S. House of Representatives. *Curtailment of Postal Services*. Washington, D.C.: U.S. GPO, 1964.

———. *Oversight Hearings on the Postal Service*. Washington, D.C.: U.S. GPO, 1973.

———. *Mail Service in Rural America*. Washington, D.C.: U.S. GPO, 1974.

———. *Postal Reorganization Act Amendments of 1975*. Washington, D.C.: U.S. GPO, 1975.
———. *GAO's Recommendation That 12,000 Small Post Offices be Closed*. Washington, D.C.: U.S. GPO, 1975.
———. *Problems of the U.S. Postal Service: A Compendium of Studies, Articles, and Statements on the U.S. Postal Service*. Washington, D.C.: U.S. GPO, 1976.
———. *Cutbacks in Postal Service*. Washington, D.C.: U.S. GPO, 1976.
———. *Postal Service Finance*. Washington, D.C.: U.S. GPO, 1976.
———. *The Postal Service Act of 1977*. Washington, D.C.: U.S. GPO, 1977.
———. *Recommendations of the Commission on Postal Service*. Washington, D.C.: U.S. GPO, 1977.
———. *General Oversight and Postal Service Budget,* Washington, D.C.: U.S. GPO, 1977
———. *Research and Development into Electronic Mail Concepts by the USPS*. Washington, D.C.: U.S. GPO, 1977
———. *Postal Research and Development*. Washington, D.C.: U.S. GPO, 1978.
U.S. Postal Rate Commission, Docket 79–6, February 2, 1979.
U.S. Postal Service. *Statutes Restricting Private Carriage of Mail and Their Administration*. House Committee on Post Office and Civil Service, 93rd Congress, 1st Session, Committee Print, Washington, D.C.: U.S. GPO, 1973.
———. *The Necessity for Change*. Washington, D.C.: U.S. GPO, 1976.
———. *Annual Report of the Postmaster General*. Washington, D.C.: U.S. GPO, 1976.
———. *Office of Advanced Mail Systems Development, Electronic Message Service System*. Washington, D.C.: U.S. GPO, 1977.
———. *Annual Report of Postmaster General* (February 1978). Washington, D.C.: U.S. GPO, 1979.
U.S. Privacy Commission. *Study*. Washington, D.C.: U.S. GPO, 1976.
U.S. Senate. *Postal Reorganization,* Part 4. Washington, D.C.: U.S. GPO, 1976.
———. *Problems of the U.S. Postal Service*. Washington, D.C.: U.S. GPO, 1976.
———. *Evaluation of the Report of the Commission on Postal Service*. Washington, D.C.: U.S. GPO, 1977.
———. *Electronic Communications and the Postal Service*. Washington, D.C.: U.S. GPO, 1977.
U.S. Senate, Committee on Labor and Public Welfare, Subcommittee on Employment, Manpower and Poverty. *Work in America,* 93rd Congress, 1st Session, February 1973.
Vezza, A. "A Model for an Electronic Postal System," in *Implications*

of Low-Cost International Non-Voice Communications, ed. I. de Sola Pool and A.B. Corte. Cambridge, Mass.: MIT Center for Policy Alternatives, September 1975.

Virginia Law Review. ''The Postal Reorganization Act: A Case Study of Regulated Industry Reform,'' *Virginia Law Review* 5, 1972, pp. 1030–98.

Wattles, G.M. ''Rates and Costs of the United States Postal Service,'' *Journal of Law and Economics* 16 (7), April 1973, pp. 89–117.

Waverman, Leonard. ''Pricing Principles: How to Price Post Office Services,'' (paper prepared for Conference on Postal Services Issues, October 13, 1978).

Wein, H.H. *Domestic Air Cargo: Its Prospects.* East Lansing, Mich.: Michigan State University, 1962.

Williams, Lawrence K., and Thomas Lodahl. ''An Opportunity for OD: The Office Revolution,'' *OD Practitioner* 10 (4), December 1978.

———. ''Comparing WP and Computers,'' *Journal of Systems Management* 29 (2), February 1978.

Withington, Frederic G. ''Future Computers and Input-Output Devices,'' *Technology Tends.* New York: IEEE Press, 1975.

Xerox Corporation. *Petition for Rule Making by FCC,* November 16, 1978.

Yankee Group. ''Making Office Automation Work,'' *Telecommunications Policy,* June 1979.

Index

About the Author

Alfred M. Lee is a postdoctoral associate holding a joint appointment with the Program on Science, Technology and Society and the Department of Environmental Engineering at Cornell University. He was educated at the University of Illinois, Urbana-Champaign, where he received the B.S. in 1973 and at Cornell University where he received the M.S. in 1975 and the Ph.D. in transportation, engineering, and public policy in 1981. Dr. Lee is the coauthor and coeditor of *Communications for a Mobile Society* (1978) and has written a number of articles dealing with telecommunications planning, technology assessment, and transportation communications interactions. He is a member of several professional societies and committees.